PATHS OF INNOVATORS

IN SCIENCE, ENGINEERING AND TECHNOLOGY

VOLUME 1

Third Edition

AF529838

R. PARTHASARATHY
Professor (Retd.), IIT Madras

Universities Press

UNIVERSITIES PRESS (INDIA) PRIVATE LIMITED

Registered Office
3-6-747/1/A & 3-6-754/1, Himayatnagar, Hyderabad 500 029 (A.P.), India
email:info@universitiespress.com, www.universitiespress.com

Distributed by
Orient Blackswan Private Limited

Registered Office
3-6-752 Himayatnagar, Hyderabad 500 029 (A.P.) India

Other Offices
Bangalore / Bhopal / Bhubaneshwar / Chandigarh / Chennai
Ernakulam / Guwahati / Hyderabad / Jaipur / Kolkata
Lucknow / Mumbai / New Delhi / Noida / Patna

© R Parthasarathy, 1999, 2003, 2012

First edition published in India by Westland Ltd, 1999
Second edition published in India by Westland Ltd, 2003
Third Edition published by Universities Press, 2012

ISBN 978-81-7371-750-5

Cover and book design:
© Universities Press (India) Private Limited 2012

All rights reserved. No part of the material may be reproduced or utilised in any form, or by any means, electronic or mechanical, including photocopying, recording, or by any information storage and retrieval system, without written permission from the publisher.

Cover design
Vinna Design Studio, Chennai

Typeset in Times New Roman 11 pt by
Trinity Designers & Typesetters, Chennai 600 041

Printed in India by
Print line Printers, Hyderabad 5000 29

Published by
Universities Press (India) Private Limited
3-6-747/1/A & 3-6-754/1, Himayatnagar, Hyderabad 500 029 (A.P.), India

---

Photographs on the cover (from top clockwise): CV Raman, Srinivasa Ramanujan, Niels Bohr, Michael Faraday, Paul Dirac, Irene Curie

Dedication

To

My Alma Mater

St. Joseph's College, Tiruchirapalli

and

Indian Institute of Science, Bangalore

FOREWORD

India needs science and technology for its development and it is necessary to attract talented young men and women into careers in science. We have to make science education interesting and science careers attractive. At the same time young persons must get acquainted not only with the work of great scientists and geniuses but also with their lives, the way they worked and succeeded, the difficulties they faced and overcame and their prejudices and failings. Of course, the layman should also learn about scientists!

The scientists selected by the author of this book span several centuries and almost all the persons who laid the foundation of modern science and built its edifice are there. Each life is so different and this comes out beautifully in this remarkable work of Dr. Parthasarathy, which is a storehouse of information and the result of almost a decade's labour. In this compilation, there were some things I knew about and many more things I did not know about. There is almost nothing in it, which is not interesting.

Charles Darwin, who developed the theory of evolution, is here. So is Charles Babbage, the originator of the concept of modern computers. Carnot, who laid the foundation of thermodynamics, Maxwell famous for his electromagnetic equations and Descartes, Pascal, Wiener, Faraday, Rutherford and Fermi — they are all here. Our own great scientists Ramanujan, Raman and Chandrasekhar are here. These are names we are familiar with but the book summarises in an easily understandable way, the highlights of their scientific contributions.

There are also fascinating snippets. Some of the scientists have had to fight great odds; Stephen Hawking, the most well-known name in modern cosmology, is crippled by the motor neuron disease and is confined to a wheel chair. Ramanujan and Leibniz taught themselves most of the mathematics they knew. Faraday and Raman were some of the greatest experimental physicists the world has known. And Hilbert and Chandrasekhar were some of the greatest teachers. Chadwick, who discovered the neutron, became a physicist by accident, attending the wrong interview. The man who discovered the electron, J.J. Thomson, was a clumsy experimentalist, though a brilliant apparatus designer and came to physics for want of money to enroll in an engineering course. Many of the illustrious careers ended very early — Carnot died at 36 and Ramanujan at 32. All these and more are in the book.

The great men had their own prejudices or were hounded by enemies. Rutherford used to say "Chemistry is a branch of physics and the rest of the sciences is stamp-collecting". Of course, today, it is certainly not true with biology moving to center stage. Raman had to resign as Director of the Indian Institute of Science after unfair attacks; he was also prevented from bringing one of the world's leading German theoretical physicists Max Born — who later won the Nobel Prize — to Bangalore by a second-rate British professor Aston who attacked Born so viciously that the latter, who was present in Bangalore when it happened, writes that "he cried" after going home. Boltzmann, whose name is associated with the constant so commonly used in thermodynamics, was so much affected by unfair criticism that he committed suicide. Eddington's vicious criticism of his work was undoubtedly responsible for the extraordinary delay in the Nobel award to Chandrasekhar. And so on.

This book can be read in any order and in parts depending on the reader's interests. It is also readable, because it is written by a physicist–engineer with an academic background and written by him on a theme he so obiously loves. This should be in every library and also in the personal collections of those who like biographies.

R. Chidambaram
Principal Scientific Advisor
Government of India

November 1, 1999

Preface to the Third Edition

In presenting this edition, revised with significant changes in the format, I pursue the same aim — to capture the love of young students and enthusiastic teachers of science and to bring out vividly the fundamental ideas the great scientists generated which got translated into inventions.

Shannon, though not distinguished at that time as a mathematician, still received the Nobel Foundation prize for the best engineering paper by an author under the age of 30. He gave shape to Information Technology (IT) by delving into Boole's classic *An Investigation of the Laws of Thought* (1854), which was gathering dust in the library.

This book is not a mere narration of facts; information hitherto unknown and not readily accessible have been brought to light. The leading strings to take you forward can be gathered from the foreword of Dr. R. Chidambaram (Principal Scientific Advisor to the Government of India).

I am grateful to Universities Press, Hyderabad, for taking up the publication of this book. My thanks are due to the Editor, Dr. Gita S Dattatri, for her active participation, the care she bestowed to the technical aspects and the splendid finish given to the work.

August 1, 2011 **R. Parthasarathy**

PREFACE TO THE FIRST EDITION

"What a piece of work is a man!
how noble in reason!
how infinite in faculty!
in action, how like an angel!
in form and moving how express and admirable!
in apprehension, how like a God!"

–Hamlet,
Act II, Scene 2

This book on some of the world's most fertile scientific minds of the past few centuries is intended as a shop-window for fundamental ideas, concepts, discoveries and techniques, all selected at random from different disciplines in science and engineering.

It is the outcome of a strong interest that I developed in the historical roots of engineering in my teaching career since the 1950s, which was spread over three premier institutions — University of Roorkee; College of Engineering, Guindy; and IIT Madras.

It seems to me this is the only way to give young students "a little foretaste of the great pleasure that awaits them in their elder years, when they come to the rich treasures from which these small and valueless coins are extracted, pretending to no other merit than imperfect stamps" — Charles and Mary Lamb, in their introduction to the *Tales from Shakespeare.*

My experience revealed to me over the years the inspiration that the magnificent lives of these great scientists and innovators would instill in young minds on the threshold of their career; or laymen interested in the history of science. It is as a consequence of this plan that nothing about their work is treated in a general way or in great depth.

So I embarked on a programme of collecting details about their lives and career quests to trace the development of their discoveries. *The Hindu* gave shape to my dream by opening up a column in its *Science and Technology* Supplement, and carrying my articles in a serial form since January 1991.

The book is organised in five parts: Engineering, Physics, Mathematics, Chemistry and Life Sciences. The men whose life sketches are traced are pathfinders in their respective fields (not confined to Nobel Laureates!). They have, the reader will notice, literally laid the foundations of modern science.

The young reader will meet a variety of personages: school dropouts, men steeped in poverty, men born to aristocracy, men battling with grave illnesses; men who changed their disciplines — from clergy to science, farming to physics, medicine to chemistry; and pairs — father and son, teacher and student, continuing the work.

This work would not have been possible but for the warm response and continued support of Mr. N. Ravi, the Editor of *The Hindu*. I wish to record my deep sense of gratitude to him.

I record my indebtedness to *The Hindu* (Prop. Kasturi & Sons Ltd), for the unstinted support to publish the serial *Paths of Innovators* from 1991 to 2006 which has culminated in the publication of this work.

It is a pleasure to acknowledge my indebtedness to Dr. R. Chidambaram, Chairman, Atomic Energy Commission Government of India, for finding the time to go through this work and for introducing it to the reader.

Suggestions for the improvement from the readers will receive careful attention.

December 1, 1999 **R Parthasarathy**

ACKNOWLEDGEMENTS

The author is indebted to the publications listed below which have been the sources of information in the preparation of the articles:

- *The Dictionary of Scientific Biography*, Ch. Scribner's Sons, New York, 1984.
- *The McGraw-Hill Encyclopedia of World Biography*, 1972.
- *The Biographical Dictionary of Scientists*, Blond Educational, London 1984.
- *The Hutchinson Dictionary of Scientific Biography*, Helicon Publishing Ltd, Oxford, 1994.
- *Biographical Dictionary of Scientists*, Harper Collins, 1994.
- *The Illustrated Science and Invention Encyclopedia*.
- *Nature* (Weekly), London.
- *New Scientist*, New Science Publications, London (Weekly).
- *Science*, American Association for the Advancement of Science, Washington D.C. (Weekly).
- *Current Science*, Indian Academy of Sciences, Bangalore; Indian National Science Academy, New Delhi.

Where appropriate, the sources are also acknowledged in the articles.

The photographs in the book have been taken from http://commons.wikimedia.org under the terms of the GNU Free Documentation License.

R. Parthasarathy

CONTENTS

Foreword v
Preface to the Third Edition vii
Preface to the First Edition viii
Acknowledgements x

SECTION ONE / ENGINEERING

Michael Faraday (1791-1867) 3
The greatest experimenter

Lord Kelvin (1824-1908) 6
Unflagging dedication to science

Clerk Maxwell (1831-1879) 9
The architect of electrical engineering

Oliver Heaviside (1850-1925) 12
The genius who lies forgotten

Heinrich Hertz (1857-1894) 15
Whose experiments led to radiotelegraphy

Jean Arago (1786-1853) 18
Who discovered the electromagnet

Joseph Henry (1797-1878) 21
Who discovered self-induction in conductors

John Baird (1888-1946) 25
Pioneer of television

Charles Steinmetz (1865-1923) 28
Who established the fundamentals of electrical power engineering .....

Nikola Tesla (1856-1943) 32
Who ushered in the power age

Geoffery Ingram Taylor (1886-1975) 35
A class academician

James Watt (1736-1819) 38
Who propelled the industrial revolution

George Stephenson (1781-1848) 42
Who invented the steam locomotive

Sadi Carnot (1796-1832) 45
Who completed the engine cycle

Robert Fulton (1765-1815) 48
Who pioneered steam navigation

Charles Babbage (1792-1871) 52
Architect of modern computers

Irving Langmuir (1881-1957) 55
Penchant for the unusual

Guglielmo Marconi (1874-1937) 57
Who invented the radio

Edward Appleton (1892-1965) 60
Pioneer of radio communication

Patrick Haggerty (1914-1980) 63
His gift for Christmas '54

Arthur Cotton (1803-1899) 67
Dedicated irrigation engineer

Alfred Krupp (1812-1887) 72
The cannon king

Hugh Callendar (1863-1930) 75
Pioneer in steam engineering

Fritz Haber (1868-1934) 78
Who invented the ammonia process

Theodore von Karman (1881-1963) 81
Who made flying faster than sound possible

Frank Whittle (1907-1996) and **Hans von Ohain** (1911-1998) 84
Who pioneered the jet engine

Karl Terzaghi (1883-1963) 87
Father of soil mechanics

Rudolf Diesel (1858-1913) 90
The man behind the 'Black Mistress'

Leo Esaki (b.1925) and **Ivar Giaever** (b.1929) 94
Pioneers in electron tunnelling and superconductivity

Robert Watson-Watt (1892-1973) 98
Pioneer of radar development

Vladimir Zworykin (1889-1982) 100
Inventor of the modern television camera

Claude Shannon (1916-2001) 103
Father of Information Technology

SECTION TWO / PHYSICS

Christian Huygens (1629-1695) 109
Who founded the wave theory of light

Henry Cavendish (1731-1810) 113
Eccentric genius

Charles Coulomb (1736-1806) 116
Who laid the basis for electromagnetics

Thomas Young (1773-1829) 119
Physician, physicist and Egyptologist

Joseph Fraunhofer (1787-1826) 122
Who identified light sources by their spectra

August Fresnel (1788-1827) 125
Who established the wave theory of light

Jean Charles Peltier (1785-1845).. 128
Who invented the reverse of Seebeck effect

Jean Foucault (1819-1868) 130
An outstanding experimental physicist

Hermann Helmholtz (1821-1894).. 134
Who laid the foundation for many a science

Josef Stefan (1835-1893) 137
Who first calculated the sun's temperature

Ernst Mach (1838-1916) 140
Who characterised supersonic flight

Ludwig Boltzmann (1844-1906).. 145
Who gave the laws of statistical physics

William Crookes (1832-1919).. 148
Who modelled himself on Faraday

Johannes van der Waals (1837-1923) 153
From school teacher to Nobel Laureate

Henri Becquerel (1852-1908) 155
Who discovered radioactivity

William Bragg (1862-1942) and **Lawrence Bragg** (1890-1971)....... 158
Like father, like son

Max Born (1882-1970).. 161
Who bridged physics and chemistry

Max Planck (1858-1947) 165
Who introduced quantisation of energy

Wilhelm Wien (1864-1928) 168
Whose law governs radiation from objects

J.J. Thomson (1856-1940) 171
Who discovered the electron

Niels Bohr (1885-1962) 174
Who modelled the atom

Robert Millikan (1868-1953) 178
Who measured the charge of the electron

Wolfgang Pauli (1900-1958) 182
Who quantised electrons

Werner Heisenberg (1901-1976) 185
Who set the limits to accuracy

Paul Dirac (1902-1984) 188
Whose passion was physics

Lord Rutherford (1871-1937).. 192
The father of nuclear science

Peter Kapitza (1894-1984) 195
Beacon of Cambridge and Moscow

James Chadwick (1891-1974) 200
Nuclear scientist by accident

Christian Doppler (1805-1853) 203
Celebrity in physics by a single discovery

William Roentgen (1845-1923) 207
The first physics Nobel Laureate

C.V. Raman (1888-1970) 210
India's first Nobel Laureate in physics

Enrico Fermi (1901-1954) 215
Who embraced all physics

SECTION THREE / MATHEMATICS

Rene Descartes (1596-1650) 221
Founder of analytical geometry

Blaise Pascal (1623-1662) 225
Mathematician, theologian and philosopher

Gottfried Leibniz (1646-1716) 229
Who influenced many sciences

Jean Argand (1768-1822) 233
Celebrity by a single invention

Friedrich Bessel (1784-1846) 236
The complete astronomer

Henrik Abel (1802-1829) 239
An immortal in group theory

Evariste Galois (1811-1832) 244
Who heralded the birth of modern algebra

George Boole (1815-1864) 247
Founder of mathematical logic

Yasudas Ramachandra (1821-1880) 250
Mathematician and educationist

Leonhard Euler (1707-1783) 254
Against all odds

Joseph Lagrange (1736-1813) 257
The self-taught genius

Pierre Laplace (1749-1827) 260
Who transformed mathematical analysis

Joseph Fourier (1768-1830) 264
Who simplified signal processing

Augustin Cauchy (1789-1857) 267
Who developed vital mathematical tools

Carl Gauss (1777-1855) 270
Austere mathematician

Bernhard Riemann (1826-1866) 273
Who gave geometry a new dimension

Edmund Halley (1656-1743) 276
The comet hunter

James Bradley (1693-1762)........................... 280
From clergy to astronomy

William Herschel (1738-1822)........................... 283
Father of stellar astronomy

John Herschel (1792-1871)........................... 287
Worthy son of an illustrious father

Henri Poincare (1854-1912)........................... 291
'Monster' of mathematics

George Cantor (1854-1918)........................... 295
Who founded the set theory

David Hilbert (1862-1943)........................... 298
A life in mathematics

Norbert Wiener (1894-1964)........................... 302
Father of cybernetics

Srinivasa Ramanujan (1887-1920)........................... 306
The greatest mathematician of India

S. Chandrasekhar (1910-1995)........................... 312
Pushed the limits of astrophysics

Paul Erdos (1913-1996)........................... 316
A homeless mathematical luminary

Stephen Hawking (b.1942)........................... 320
Who works amidst formidable disability

SECTION FOUR / CHEMISTRY

Joseph Black (1728-1799)........................... 327
Founder of calorimetry

Antoine Fourcroy (1755-1809)........................... 331
From medicine to applications of chemistry

Antoine Lavoisier (1743-1794)........................... 334
Founder of modern chemistry

Claude Berthollet (1748-1822)........................... 337
Who founded applied chemistry

John Dalton (1766-1844)........................... 341
Who charted the course of chemistry

Amedeo Avogadro (1776-1856)........................... 344
From law to physics and chemistry

Joseph Gay-Lussac (1778-1850)........................... 348
Scientist who worked with his hands

Jacob Berzelius (1779-1848)........................... 352
From medicine to chemistry

Pierre Dulong (1785-1838)........................... 356
Who founded thermochemistry

Justus von Liebig (1803-1873)........................... 360
Who took chemistry to students

August von Hofmann (1818-1892)........................... 363
Who heralded German chemical industry

Friederich Kekule (1829-1896) 366
Who classified organic compounds

Robert Bunsen (1811-1889) 369
Of burner fame

Dmitri Mendeleev (1834-1907) 373
Who grouped the elements

William Perkin (1838-1907) 375
Who discovered synthetic dyes

William Ramsay (1852-1916) 378
Who discovered inert gases

Svante Arrhenius (1859-1927) 381
Pioneer of modern physical chemistry

Morris Travers (1872-1961) 384
Who discovered the missing elements

Irene Joliot-Curie (1897-1956) 388
Discoverer of artificial radioactivity

Dorothy Hodgkin (1910-1994) 391
Chemistry's crystal seer

SECTION FIVE / LIFE SCIENCE

William Harvey (1578-1657) 397
Founder of modern experimental physiology

Antony von Leuwenhoek (1632-1723) 401
Hunter for microbes

Albrecht von Haller (1708-1777) 406
Anatomist turned science writer

Friedrich Woehler (1800-1882) 409
Pioneer of biochemistry

Charles Darwin (1809-1882) 412
Who revolutionised biology

Alfred Wallace (1823-1913) 417
A dedicated naturalist

Ivan Pavlov (1849-1936) 419
Pioneer in physiology

Paul Ehrlich (1854-1915) 423
The founder of chemotherapy

Sigmund Freud (1856-1939) 426
Founder of psychoanalysis

Willem Einthoven (1860-1927) 431
Inventor of the electrocardiograph (ECG)

Salim Ali (1896-1974) 433
Bird lover and ecologist

G.N. Ramachandran (1922-2001) 437
A pioneer in molecular biophysics

Index 440

Maxwell

SECTION ONE

Whittle

ENGINEERING

Esaki

Michael Faraday (1791-1867)

The greatest experimenter

Great men of science have through the centuries faced the layman's question: ***'What is the practical use of their discoveries?'*** Faraday gave a demonstration of the phenomenon of electromagnetic induction in 1831, when one in the audience peremptorily asked him, ***"The behaviour of the magnet and the coil is interesting, but of what possible use can it be?"*** We now witness that Faraday's demonstration has grown into the giant electrical industry — the modern electric motor, generator and transformer, which are instantly recognisable today.

Crisis in boyhood

Born on September 22, 1791 into a poor family, which migrated to London from the north of England, Michael Faraday received the bare rudiments of an education — the three R's in a Sunday church school. His mother gave him support emotionally, with ***'a loaf of bread that had to last him for a week.'***

An event that changed his career

At the age of 14, Faraday began to earn his livelihood by delivering newspapers and later worked as an apprentice to a bookbinder. Mr. Riebau, a gentleman calling one day at his master's shop, found Faraday pouring over the article 'Electricity' in an encyclopedia placed in his hand for binding. He gave the young bookbinder, tickets to attend at the Royal Institution a course of four lectures delivered by Sir Humphrey Davy (1778-1829). ***The impact on Faraday was profound and changed the course of his career.***

Faraday had written out neatly the lectures given by Davy, illustrated them with drawings; this notebook he sent to Davy, with a request for employment in his laboratory. Faraday spent his whole working career in the Royal Institution (1813-1858), first as Davy's assistant and finally as his scientific successor as Director of the laboratory.

Michael Faraday is considered ***the greatest experimenter of all time.*** Among his many discoveries, the following are significant:

- The law of electromagnetic induction, which laid the foundation for the electrical industry.
- The conception of electric lines of force, to which Maxwell (1831-1879) gave a mathematical form, 30 years later.
- The laws of electrolysis, which form the basis for electrochemistry. another big modern industry.
- Liquefication of chlorine.
- Isolation of benzene.
- The rotation of plane of polarisation of a light beam by a magnetic field.
- Diamagnetism and the magnetic properties of all matter.

Faraday ***contributed many key electrical terms*** to the English language — electrolyte, electrolysis, electrode, cathode, anode and ion, all used for the first time in a paper he published in 1834.

Breakdown in health

The succession of these extraordinary discoveries through innumerable experiments — requiring continual modification of ideas — drove Faraday to the limits of physical exhaustion. He suffered a severe health breakdown in 1839 and there was little creative output for the following five years. After reviving, a series of new discoveries and important contributions followed in the time from his 53rd to 64th year; in 1858, Faraday retired from the Royal Institution and moved to a house outside London presented to him by the Queen. He died on August 25, 1867, at the age of 76. He was elected FRS in 1824.

A consummate lecturer

Faraday was an admired successor of Davy too, in the delivery of the ***Friday evening lectures at the Royal Institution.*** This success had not come without its share of hard work. Not sure if he could present his ideas in a lucid and congenial manner, Faraday educated himself in 'English composition, style, grammar, delivery and oratory'. The hard-won practical wisdom has been made available to others in the anthology ***'Michael Faraday's advice to a lecturer' published by the Royal Institution in 1974.***

Faraday's strength of character was derived from a deeply religious foundation, the small sect of the Sandemenians. One of their tenets was that it was improper for any member to accumulate wealth, which may possibly explain why he did not have any of his discoveries patented.

Devotion to science

Faraday lived on a high level of thought and feeling: he shunned publicity, as it was an unwelcome deviation from his complete devotion to the pursuit of science. It was for this reason that he declined to become the President of the Royal Institution. Likewise, ***he rejected knighthood: he would, he said, remain plain Michael Faraday*** to the end. A £20 note was released by the Bank of England in U.K. in honour of Faraday.

Lord Kelvin (1824-1908)

Unflagging dedication to science

Together with Helmholtz, he occupied a pre-eminent position in creating the 'Science of Physics', as it was known in 1900.

Educated at Glasgow and Cambridge

William Thomson (who became Lord Kelvin in 1892) was born at Belfast in Ireland on June 26, 1824. His father was a teacher of mathematics and the author of several noted texts on calculus; he undertook the entire education of his son up to his tenth year, and in a great many directions too.

William was educated at Glasgow and Cambridge Universities. His studies at Cambridge (1840-45) did not influence him as deeply as his years at Glasgow did, because of the extreme importance attached to finishing the examinations in the first rank and passing out as 'senior wrangler'. However, while in Cambridge, he had already made a name in mathematical physics through original publications on Fourier's theory of heat conduction.

Soon after his graduation in 1845, at his father's suggestion, he went to Paris and worked for a year in Regnault's Laboratory. There he made the acquaintances of eminent French mathematicians, Cauchy and Sturm.

Passion for teaching

Shortly after his return to Glasgow in 1846, he was appointed Professor of natural philosophy at the incredibly young age of 22. He held this position at the Glasgow University for the rest of his life, combining remarkably scientific work with a stimulating degree of teaching activity. He laid out his laboratory in an empty wine cellar and was soon able to augment the very modest means at his disposal through his technical inventions. He was knighted in 1867 for leading ***the engineering project, which made possible the telegraphic cable communication*** between England and the U.S. He died on December 17, 1907.

At the age of 75, he gave up teaching; he had such passion for this activity that he recalled after his retirement, how "My predecessor in the natural philosophy chair taught his students reverence for the great French mathematicians Legendre, Lagrange and Laplace. My successor added Fresenl and Fourier to this list of scientific nobles".

Contributions

Kelvin's contributions to science and engineering were phenomenal: these include, apart from the absolute scale of temperature, in kelvin, ***the second law of thermodynamics,*** formulation of the theories of electric and magnetic fields, invention of the quadrant electrometer and the mirror galvanometer, the receiving apparatus for cable telegraphy and the telegraph equation and his work on physical geology.

Kelvin was the first to prove how the self-induction of a circuit, discovered by Faraday (1791-1867), must result in the production of alternating electric currents in an open circuit of conductors, and to calculate every detail of electric oscillations, namely their period, damping. Maxwell (1831-79) added the idea of electric waves, which should emanate from such oscillations, and it was left to Hertz (1857-94) later to demonstrate the production of these waves.

Towards the end of his life, Kelvin's reputation suffered a slight setback when he led a losing battle against Rutherford's (1871-1937) theory of radioactivity and also opposed Maxwell's electromagnetic theory in favour of more elaborate and less successful models.

Creating the science of physics

However, it is acknowledged that, together with Helmholtz (1821-94), he occupied a pre-eminent position in creating the science of physics, as it was known in 1900. The absolute scale of temperature has been named after him.

Lord Kelvin is remembered for his two oft-quoted statements, which embody ***practical advice to men embarking on a career in science:***

i. 'You can't really understand something unless you can make a mechanical model of it.'
ii. 'When you can measure what you are speaking about and express it in numbers, you know something about it. But when you cannot, you have scarcely, in your thoughts, advanced to the stage of a science, whatever the matter may be.'

Clerk Maxwell (1831-1879)

The architect of electrical engineering

Electromagnetism is the phenomenon that gives light for illumination, power for driving appliances, radio for broadcasting and communication, television for entertainment and a host of other building blocks for our technological civilisation. ***Electricity and magnetism are so closely related that, as an old song has it, "you can't have one without the other".*** It is Clerk Maxwell who combined the then existing laws of electricity and magnetism — uncovered in isolated bits during the previous hundred years, into differential equations of great power and beauty. He proved mathematically in 1865 that the electromagnetic phenomena are propagated as ***waves through space with the velocity of light.***

Early years

Maxwell was born in Scotland in 1831 and had his early education in the University of Edinburgh, which he entered at the age of 13. In 1847, he proceeded to Cambridge where he distinguished himself by his research publications in mathematics and physics. He was appointed professor of physics in Aberdeen at the young age of 25. He shifted to the King's College, London in 1860; it was here that he met Faraday (1791-1867).

In 1865, ***he retired to his Scottish estate where he devoted his time entirely to scientific work***. The result was the two volumes of his masterpiece 'Treatise on Electricity and Magnetism' (1873). The special features of

Maxwell's theoretical work had, however, to wait for their recognition till the experimental discoveries of Hertz (1857-1994) in 1888.

Return to academic activity

Maxwell returned to academic life in 1871 and succeeded Lord Rayleigh (1842-1919) as Professor of experimental physics at Cambridge. He built and supervised the Cavendish Laboratory; besides he edited the monumental work 'Electrical Researches of Cavendish'. He somehow found the time and took delight in serving as examiner for the Mathematical Tripos. He instituted reforms in the substance and style of the examinations, which were widely acclaimed in the academic world.

Only eight years later, in his 48th year (1879), a sudden illness cut his life short. It is to be noted that Maxwell did not arrive at his equations directly by mathematical summarisation but from Faraday's experimental investigations. He himself said that he put Faraday's ideas into a mathematical form.

The power of his equations

Maxwell's equations summarise, in an astonishing form, the fundamental results of the whole series of researches done earlier: ***they contain Coulomb's law, Gauss's law, Ampere's law and Faraday's law of induction*** — all this being supported on Faraday's conception of lines of force.

As regards their power of application, it may be added that Maxwell's work resulted in Faraday's idea being accepted by technical workers. This was all the more important, as Faraday's induction had found increasing application in the production of strong electric currents, the replacement of permanent magnets by electromagnets.

The unnecessary long limbs of the electromagnets disappeared and the modern form, most favorable to the development of lines of force, brought with it the high efficiency obtainable today in the transformation of mechanical into electrical energy and vice versa. ***The foundations of electrical engineering were thus laid.***

Other contributions

Maxwell's contributions were prolific and diverse, arising out of his investigations on:

- kinetic theory of gases,
- statistical approach of physical phenomena,
- photo elasticity,
- colour vision,
- mathematical description of control system and
- the stability of Saturn's rings.

It is now established that the greatest of all advances in physics made in the 19th Century was by Clerk Maxwell. ***"The epic electromagnetic equations that bear his name constitute one of the greatest creative achievements of humans*** comparable only with that of Newton in dynamics" (Centennial Publication of The Institution of Electrical and Electronic Engineers, New york).

"His name stands magnificiently over the portal of classical physics, and we can say this of him: by his birth Clerk Maxwell belongs to Edinburgh, by his personality he belongs to Cambridge, ***by his work he belongs to the whole world***." *(Max Planck)*

Oliver Heaviside (1850-1925)

The genius who lies forgotten

Today, it is taken for granted that telephone transmissions will be clear. Although transcontinental communication is now by satellite, it was Heaviside who revolutionised the use of telephone by establishing the condition for distortionless signal transmission. This breakthrough was achieved by using mathematics as the only tool.

Early life

Oliver Heaviside was born in London on May 18, 1850. He had a meager education and discontinued school at the age of 16. In later years, he would recall that ***the first 13 years of his life in poverty permanently deformed his future***. The family could provide him one signal benefit: one of the members, Sir Charles Wheatstone (1802-1875), encouraged the young Heaviside's life-long devotion to the study of electricity and helped him to secure a job as a telegraph operator.

From 1868 to 1874, he did this job well and he managed to study, in his spare time, language, mathematics, electricity and magnetism. He thus prepared himself for his future work, though at that time he showed no promise of any analytical abilities. ***He led a hermit-like existence,*** often in desperate financial straits.

Amongst ***his contributions to modern electrical engineering,*** the following are outstanding.

- The mathematical condition for distortionless telephone lines and the addition of inductive loading coils to telephone lines.
- The use of operational calculus to tackle physical problems involving discontinuities.
- ***The formulation of Maxwell's field equations in the modern form.***
- The existence of a conducting layer in the atmosphere aiding the transmission of radio waves (Heaviside layer).

For none of these achievements has Heaviside received his due recognition today. A search for his name in textbooks will show only a brief mention. Others received the credit for his inventions. Besides his original contributions, ***Heaviside enriched the language of electrical engineering.*** Lamenting the lack of a proper name for the a.c. resistance of inductive circuits, Lord Rayleigh remarked in a paper (1886) that Heaviside's term was 'impedance', now a common terminology in electrical engineering.

Clash with the post office

Many of his contemporaries with university education acknowledged Heaviside's contributions. But the influential person in the British post office, Sir William Preece was his nemesis. Heaviside's radically new idea of adding inductance to telephone lines was simply beyond Preece's ability to grasp: there was hence a monumental clash between these two men. The Royal Society mathematicians too had no sympathy, because Heaviside got the right answers to problems no one could solve. They considered his methods mathematical blasphemy.

Amidst such a background, it is not surprising that Heaviside decided to opt for early retirement in 1874, when he was just 24. But at what incredible sacrifice, embracing ***a life of quiet poverty for the pursuit of scientific work!***

Apostle of Maxwell

Heaviside commenced his intense study of Maxwell's 'Treatise on Electricity and Magnetism', which provided him the required tools to investigate the problems in telephone transmission lines. He was literally, an apostle of

Maxwell's (1831-18-19), saying, "I preach the gospel". ***He cleared the maze of symbols in Maxwell's theory and reduced them to two, electric and magnetic vectors.***

Kelvin thought electricity travelled through a wire by the process of diffusion, much like heat. Heaviside's profound research established the theory of electrical wave propagation in conductors. By his mathematical theory he showed conclusively that 'electromagnetic induction is a positive benefit; it helps to carry the current'. Lord Kelvin acknowledged ***this work on submarine cable analysis as one instance of the highest mathematical power working and judging physical applications.***

Heaviside published his work in *The Electrician.* His articles were difficult to grasp, as he often used his own highly developed intuition in place of rigorous proofs. With his operational calculus, using p and $1/p$ to denote the operation of differentiation and integration respectively, the results involved the square root of p. He boldly pushed on and got answers to real physical problems. His motto had been, "First get on, in any way possible, and let the logic be left for later work".

Honours

The IEE, London elected him an Honorary Fellow in 1918 and honoured him as ***the first recipient of its Faraday Medal.*** But even as he shivered from lack of heat, he kept his sense of humour and self-worth. When a kindly neighbour saw him sitting cold and ill in his garden, she told him to go inside and sit by the fire. "Madam", replied he, "I have no fire — I have only my genius to keep me warm".

Oliver Heaviside died on February 3, 1925, almost unnoticed. He is forgotten today, "because we fail to appreciate the agony and tears that accompany discoveries now routinely discussed in undergraduate textbooks. In our age of satellite communications, which have made the transoceanic cables of Kelvin and Heaviside obsolete, Heaviside's mathematics continues to survive as a tool in the search for even better technology". (*IEEE Spectrum*, New York)

Heinrich Hertz (1857-1894)

Whose experiments led to radiotelegraphy

Today when people listen to a radio receiver, watch a television programme of travel in an aircraft guided by radar, most may not be aware that they are enjoying the benefits of Heinrich Hertz's work. Since the enunciation of Maxwell's theory for electromagnetic fields in 1873, experimental proof was lacking. It fell to Hermann Helmholtz (1821-94) to set this problem, as a prize question of the Berlin Academy, to Hertz who enrolled as a student in his physical laboratory in 1878. He worked on this but achieved success only in Kalsruhe.

Changeover to physics

Hertz was born on 1857 at Hamburg (Germany) and attended high school till 1875 in his native town. He first chose to study engineering, because his modesty seemed to have made him doubtful of his talent for theoretical physics. Perhaps the prevailing atmosphere in his native town influenced him, which encouraged interest in practical work. But in 1877, he abandoned engineering in favour of physics, for as he acquired knowledge he became convinced that ***only the pursuit of science could give him enduring satisfaction.***

At school, he was a bright pupil, a linguist and a devoted student of literature and philosophy. He would recite Homer or the Greek tragedies freely, and turned seriously to the study of Arabic and Sanskrit for a while.

In 1878, Hertz became an assistant to Helmholtz in Berlin before moving to Kiel University three years later; two years afterward, he was appointed Professor of Physics at the University of Kalsruhe. From 1885, he began publishing a number of papers and made his most important discoveries on electric waves. Here he met Elizabeth Doll, the daughter of a colleague, whom he married in 1885. In 1889, he moved to the University of Bonn, succeeding Clausius (1822-88) the famous physicist. Shortly after, he was afflicted with an intractable jaw infection; amidst intense suffering, he found consolation in publishing his remarkable book *Principles of Mechanics.* Five years later, he died at the age of 37.

Experimental demonstration

Let us look at his original investigations where he gave spectacular proof that electromagnetic effects propagate as waves in a medium. Hertz experimented with an induction coil containing a spark gap and noticed that each spark discharge also caused a spark across a second coil nearby. The effect was at first feeble and irregular and by elaborating the circuit, he demonstrated the operation of what turned out to be a transmitter and a receiver.

He then gradually moved the receiver away until the spark grew fainter and finally went out altogether. He made it reappear by interposing slabs of various dielectrics such as paraffin. The impact of this crucial experiment is better given in the words of his mentor, Helmholtz:

"In the whole investigation ***one scarcely knows which to admire most, his experimental skill or the acuteness of his observation,*** so happily the two are in combination".

Helmholtz's tribute

Hertzian waves were the subject of experimental research from 1890 onwards, in physics laboratories of several countries in the continent. Hertz, being too ill, no longer participated. His assistant, Philipp Lenard (1862-1947), carried on most of his work and completed an experiment that fetched the Nobel Prize.

The paper describing this work was Hertz's last contribution. ***The first practical application of Hertz's work was the development of radiotelegraphy, which followed immediately.***

In a moving tribute to his student, Helmholtz wrote:

"Amongst all my pupils, I have ever regarded Hertz as the one who had penetrated further into my own circle of scientific thought and it was to him that I looked with confidence for the further development of my work. Hertz seemed to be predestined ***to open up to mankind many of the secrets which nature has hitherto concealed from us."***

JEAN ARAGO (1786-1853)

WHO DISCOVERED THE ELECTROMAGNET

On September 20, 1820, Arago announced the discovery of temporary magnetisation of soft iron by an electric current. This invention provided the technological key to the electric telegraph.

Education and career

Jean Arago was born on February 26, 1786 at Estagel, France. After completing the traditional classical education, he entered the elite Ecole Polytechnique, Paris, by studying the works of mathematicians, Euler (1701-83), Lagrange (1736-1813) and Laplace (1749-1821). He was admitted to the institution in 1803 and, while still a student in the course, he was named secretary of the Bereau des Longitudes. He was sent to Spain in 1805 along with the reputed physicist Biot on a geodetic expedition. On returning home in June 1809, he was appointed professor at the Ecole Polytechnique. Here, he taught a variety of subjects until 1830.

Arago was drawn into politics following the French Revolution. He was repeatedly elected between 1830 and 1852 to the Chamber of Deputies, where he delivered influential speeches on educational reform, application of scientific knowledge and on technical subjects such as railroads, canals and the electric telegraph. His political career reached its peak in February 1848, when he was made a minister of the provisional Government; it ended abruptly in June when he lost his control over the revolutionaries.

Theoretical contributions

Arago's original work in science was done before he entered politics, before his forty-fourth year. He achieved fundamental results in optics, electricity and magnetism. He initially worked on the refraction of light passing through the earth's atmosphere. He verified the formulae given by Laplace in his *'Celestial Mechanics',* which assumed that the atmosphere consisted of concentric rings of a mixture of oxygen and nitrogen, density varying as a function of height.

As a member of the Bureau des Longitudes, Arago made theoretical contributions, important among them being the explanation about the scintillation of stars using interference. But the stimulus he gave to young astronomers — Paul Langevin, Herve Faye and Charles Mathieu — turned him into a reputed astronomer.

Defender of wave theory

Extending the work in Thomas Young's classic paper (1801) and working on the principle of polarisation discovered by Malus (1808), Arago obtained important results: verification of the wave theory of light, elaboration of the conditions for obtaining Newton's rings, discovery of chromatic polarisation and special cases of rotary polarisation (1812). With Fresnel (1788-1827), he published a series of papers on the polarisation of light, establishing the fundamental laws of polarisation and answering several criticisms of his colleagues including Laplace and Blot.

It was Arago who suggested the crucial experiment to decide between corpuscular and wave theories of light by comparing the speed of light in water and air. The experiment, which was carried out by Foucault (1819-68) and announced in Arago's presence to the French Academy in 1850, vindicated the wave theory. Arago became a public defender of the new theory.

Magnetism of soft iron

On September 20, 1820, he announced the discovery of temporary magnetisation of soft iron by an electric current, which heralded great technological impact. This invention helped in making stronger magnets (than the conventional ones made by a loadstone). ***The electromagnet is now used***

in all appliances, from the electric bell to the giant electric locomotive. It also provided the technological key to the electric telegraph.

Other discoveries

Arago took up the study of optical instruments that proved useful in physics, meteorology and astronomy. He built (1811) the polariscope to determine the degree of polarisation of light rays. This was converted into a polarimeter, which extended the range of application: for example, to differentiate between light emanating from solid and liquid surfaces and to calculate the height of isolated clouds. He invented the cyanometer to measure the blues of the atmosphere, which was later adapted for determining the depth of the sea.

Discovery of Arago's disc

In 1822, while making certain magnetic measurements at Greenwich, he causally noticed the dampening effect that metallic substances had on the oscillations of the compass needle. After a delay of several years during which he worked on the chemical and thermal effects of light, he recognised the value of his earlier observation at Greenwich. He pursued and discovered that the rotation of a non-magnetic metallic substance (copper) created a magnetic effect on a magnetised needle (known as 'Arago's disc'). This discovery won him in 1825 the Royal Society's Copley Medal. Faraday later supported the phenomenon by his theory of induction (1831).

Public lectures

Arago enjoyed life-long friendship with Gay-Lussac and Alexander von Humboldt. He and Ampere (1775-1836) often worked in each other's laboratories. He gave popular public lectures at the Paris Observatory from 1813 to 1846. In both his writings and public appearances, Arago adopted a personal style, inquisitive, volatile and constantly bubbling with optimism. This won him a large following by devoted students.

Later he gradually lost his eyesight, but not his mental energies. His passion for physics conveying a contagious sense of excitement that spilled over into his work habits. He died in Paris on October 2, 1853.

Joseph Henry (1797-1878)

Who discovered self-induction in conductors

Henry's success in forming the Smithsonian Institution according to his ideals established his astuteness as an administrator and his pre-eminence in the American scientific community.

Early years

Joseph Henry was born on December 17, 1797 at Albany near New York, in a poor family of Scottish origin. He attended the local schools; in 1811, on the death of his father, he shifted to New York to reside with his relatives. He underwent a few years later, an apprenticeship under a local watchmaker and also a silversmith. ***George Gregory's book*** *Popular Lectures on Experimental Philosophy, Astronomy and Chemistry,* which he came across by chance, ***turned his interest to science***.

Struggle to develop himself in science

From 1819 to 1822, Henry was enrolled in the Albany Academy studying integral calculus, chemistry and natural philosophy. His main problem was how to support himself in his venture to develop as a scientist. ***So he took up odd jobs to raise the finance:*** first as a teacher in a rural school, assistant in the chemistry laboratory (1823-24), tutor in the household of a rich family, tutor to the elder Henry James in the Presidential household, and head of a survey party for a road from Hudson River to Lake Eire. In 1826, the

Albany Academy appointed Henry as professor of mathematics and natural philosophy, which post he held till October 1832. Then he became chairman at the College of New Jersey (now Princeton University) where he served till 1846.

The period he spent at the Albany Academy was indeed formative for his future scientific work. He read avidly the works in science, the journals and monographs available in the library, since he was the librarian of both the Academy and the Albany Institute of History and Art.

By 1827, when he started his work in electricity and magnetism, he was familiar with the work of leading scientists like Ampere (1775-1836), Davy (1718-1829), Faraday (1791-1867) and Young (1773-1829). Besides, he read and annotated Biot's work (in French) on electromagnetism.

Construction of powerful electromagnets

Henry's work at Albany led to the independent discovery of electromagnetic induction and to the development of powerful electromagnets. This, he avowed in his lectures at Princeton, was an application of Ampere's theories.

From Ampere's picture of the earth as a great voltaic pile with innumerable layers of materials producing circular currents around the magnetics axis, Henry probably got the concept of winding his horseshoe magnet using strands of wire, instead of using one strand which was the practice followed at that time.

Henry continued at Princeton his earlier Albany investigations; there was visibly a boost to his interests in electricity and magnetism. He was an admirer of his predecessor Benjamin Franklin (1706-1790) even to the extent of using the pseudonym 'F'. He showed particular concern for integrating the static electricity phenomena, earlier studied by Cavendish (1731-1810) and Poisson (1781-1840), with the most recent galvanic experiments.

Meeting Faraday

Henry designed devices suitable for classroom demonstrations, namely to get large effects from small inputs, whereas Faraday devised ingenious experimental set-ups to detect small effects. This is the difference in the approach between Henry and Faraday as experimenters.

Faraday invited Henry to lecture in 1837 at the Royal Institution on 'the mathematical theory of electromagnetism'. Unlike Faraday, ***Henry possessed good mathematical skill and was able to express experimental findings into mathematical formulations.*** However, he could not accept Faraday's field concept because of his staunch Newtonian beliefs.

From 1832 to 1837 (when he undertook his first European trip), there was a decline in his research output due to the pressure of teaching activities at Princeton. But from 1838, his research activities increased covering areas outside electricity and magnetism, which are not so well known.

Henry pursued studies in terrestrial magnetism, meteorology and geophysics. He published papers on phosphorescence (1841), capillarity (1845), atomicity (1846), and the theory of the imponderables (1859). He carried out considerable experimental work on the propagation and detection of light and sound. He published several papers on aurora and on heat.

Developing the Smithsonian institution

Henry's most important contribution to science was made after he took charge in 1846 as secretary of the Smithsonian Institution. There was utter confusion of aims over the bequest when the founder, the natural son of an English duke, dedicated in his will his estate to an institution in Washington D.C. for the "increase and diffusion of knowledge among men".

Henry attempted to give shape to the concept of a 'scientific community' at a time when it was little understood by many of his contemporaries: a small group of dedicated men meeting internationally recognised standards.

Henry did not occupy himself with popularisation of science or with education but with support to research and dissemination of its findings. He established an international system for exchange of scientific publications, which led in 1855 to the publication of the *Royal Society of Scientific Papers.* He saw scientific research neglected in America in favour of other human activities and so justified support of research and scholarly publications as better use of the modest size of the bequest.

Henry regarded natural selection as the best chance yet to give natural history the rigor it had lacked, after Charles Darwin published his celebrated work *The Origin of Species.*

His ideals

Henry laid the principle, which was against Smithsonian involvement in applied research, by which he was upholding the anterior of pure science. For example, in open public controversy with Morse over the telegraph, he was protecting the role of disinterested scientific research over investigations of specific practical problems. Thus his success in forming the Smithsonian according to his ideals established his astuteness as an administrator and his pre-eminence in the American scientific community.

Henry always referred to himself as self-taught. His activity in science continued throughout his long life, which ended on May 13, 1878. ***He independently discovered self-induction.*** This discovery reported in his 1835 paper, while fitting his theoretical views, was not particularly crucial to Faraday's concepts. ***His name is immortalised in electrical engineering, the unit of inductance being called 'henry'.*** (*The Dictionary of Scientific Biography,* Ch. Scribner's Sons, New York)

John Baird (1888-1946)

Pioneer of television

After two years' research, he contrived an electromechanical system consisting of a camera behind a rotating disc. He demonstrated television by telephone line between London and Glasgow.

Early life

John Baird was born on August 13, 1888, in Helensburg, Scotland. He went to school at Larchfield, received a diploma in electrical engineering at the Royal Technical College and studied for his B.Sc degree at the University of Glasgow. His studies were interrupted by the outbreak of World War I.

Baird entered in 1914 the military service in a temporary post as an electrical engineer. Persistent ill health caused him to resign the post that he held during the war years. He had to abandon three successive business ventures in Glasgow, the West Indies and London.

In these unpromising circumstances, he turned to amateur investigations, first at making artificial diamonds in which he failed.

Developing Noctovision

In 1922, he retired after a complete breakdown to live at Hastings. He was aware that different forms of electromagnetic radiation, of which visible light was only one, might be used to illuminate a person or object being televised. He soon discovered that infrared radiation was almost as effective as light.

Baird developed a ***system for transmitting pictures by the use of infrared rays.*** He called it 'Noctovision' and the receiver 'Noctovisior', since it was a night seeing eye.

He explored also the use of very short radio waves to illuminate objects. He published a paper 'An account of some experiments in television' in *The Wireless World* and *Radio Review,* May 1924. On October 30, 1925, Baird was successful in transmitting his first picture, the head of a dummy. He was keen to obtain publicity and visited the office of the *Daily Express* newspaper.

The news editor was terrified, as one of his staff reported, "For God's sake, go down to the reception and get rid of a lunatic who is down there. ***He says he has got a machine for seeing by wireless***!"

After two years' research, he contrived a primitive apparatus: his electromechanical system consisted of a light sensitive camera behind a rotating disc.

Uneven image

It delivered a crude picture consisting of 30 lines at twelve frames per second to a television receiver that displayed an uneven and tiny orange and black image.

In 1927, Baird demonstrated television by telephone line between London and Glasgow. He formed the Baird Television Development Company, which, on 30 September 1929, was allowed to provide the first BBC television program.

Synchronisation of sound and vision was achieved a few months later, and half-hour programs were shown regularly on five mornings a week. These acquired considerable publicity, when they screened the 1931 Derby.

Appleton's advice

The first formal report of the ability of a television signal to provide the range of signals reflected from an object was published by Professor Edward Appleton in *Wireless World,* January 1931. Though his main research was devoted to the study of the ionosphere, he was a part-time consultant to Baird Television Co. He advised the company on employing cathode ray tubes and ultra-short waves. So the Baird Company started experimental ultra-short wave television transmission in April 1932.

A self-taught innovator

In January 1935, the 30-line pictures of primitive television were discontinued. For two years, Braid's company operated side by side with its rival, which had the resource of the Marconi Organisation behind it.

Baird's system, working on 240 lines, mechanically scanned, and then finally gave place to the 450-line electronically scanned system promoted by Marconi-EMI.

Baird was described as a self-taught innovator ***who matched inventive wits against the pooled ability and the resources of great laboratory physicists.***

He was at the forefront of virtually all developments in television and continued research into colour; stereoscopic and big screen television until his death (June 14,1946). (*Biographical Dictionary of Scientists,* Harper Collins, 1994)

Charles Steinmetz (1865-1923)

Who established the fundamentals of electrical power engineering

The law governing magnetic hysterisis that he enunciated did for magnetic circuits what Ohm's law did for electric circuits. The constant that he calculated for the hysterisis loss has been named after him.

Migration to the U.S.

Steinmetz was born on April 9, 1865 in Breslau, Germany. At birth, he was afflicted with a hereditary physical deformity — hunchback. In the local gymnasium, he displayed a passion for mathematics, physics and literature. He entered in 1882 the University of Breslau, where he studied for five years. As he pursued his education, he also continued his political activities as an ardent socialist, a pattern he was to continue throughout his life. He was placed in 1887 under police surveillance for a stirring article he contributed to *The People's Voice.* Just then he had finished his work for a Ph.D in mathematics but ***had to flee to Switzerland*** (1888) to escape arrest. ***He never received his degree.***

Steinmetz found it necessary for both economic and political reasons to immigrate to the U.S. (1889) ***"without a penny in his pocket and not a word of English".*** He became in 1894 an American citizen whereupon he took the name Charles Proteus. He was employed as a draftsman by Rudolf Eickemeyer, who was a flourishing manufacturer of electrical machinery and a pioneer in electrical research.

The obstacles in his career

He had to overcome many obstacles — his lack of English, his physical deformity and his unfamiliarity with American professional practice. He began at once to tackle some of the mathematical problems posed by alternating current (a.c.) circuits and established himself in a short span of four years, as ***one of the most creative mathematical minds in electrical engineering***. He demonstrated the profitability of applying sophisticated mathematical techniques to practical problems.

In 1893, the newly formed General Electric purchased Eickemeyer's company, along with the services of Steinmetz; this set ***his career in a path for research and innovation.*** There Steinmetz began to indoctrinate the engineers with his method of calculating a.c. circuits.

Law for calculating hysteresis loss

In 1892 Steinmetz read, before the American Institute of Electrical Engineers, two long papers in which he presented his experiments on power losses in magnetic materials. The law governing magnetic hysteresis that he enunciated, did for magnetic circuits what Ohm's law did for electric circuits, The constant that he calculated for hysteresis loss has been named after him. ***With Steinmetz's law, the losses in the electrical machines could be calculated in advance and minimised.***

Use of complex numbers

Steinmetz's second contribution was induction of methods using complex numbers to solve practical problems that occurred in ***the design of electrical machinery and power system.*** By his method, the performance could be predicted without the necessity of the expensive process of building the system first and then testing it for its efficiency. His method of calculation was presented at the International Electrical Congress, Chicago in 1893.

Interest in electrical engineering education

Steinmetz had an ***abiding interest in education.*** He taught for over ten years in Union College (1903-1913), where he created the electrical engineering department. ***He established classes to take special care of subnormal anemic and handicapped children.*** His first textbook on electrical engineering *Theory and Calculation of Alternating Current Phenomena* (1897) was read and understood only by a few. The lack of support that he faced was due to the fact that ***electrical engineers were not taught enough mathematics to understand his mathematical treatment.*** So Steinmetz wrote a book *Engineering Mathematics,* to educate the electrical engineering profession and followed his campaign by writing and lecturing. Most of his ten technical books were received well in colleges and had a tremendous influence on curriculum development. A century later, electrical engineering students still learn and use his methods.

Steinmetz's third major technical achievement was in the ***study of transient phenomena*** in electrical circuits. A prime example of this phenomenon, lightning, attracted his attention as a mathematical problem. He designed and built a high voltage 'artificial lightning' testing laboratory in the General Electric, where he remained till the end of his life on October 26, 1923. The giant sparks crackling across the laboratory had a dramatic effect upon a public already convinced that science was a fertile source of miracles.

Other contributions

His wide-ranging contributions can be gleaned from the following list: two-phase to three-phase transformation, design of high voltage a.c. transmission techniques, mercury-arc lighting, improved design of motors, generators and transformers, almost 200 patents for useful devices and equipment that propelled the expansion of electric power systems.

Despite his top management post in a giant business corporation, he retained a lifelong interest in socialism and was one of the country's most famous socialist politicians and writer. (*STEINMETZ: Engineer and Socialist,* Ronald R. Kline, John Hopkins University Press, Baltimore, USA, 1992. Studies in the History of Technology)

Honours

Steinmetz won several laurels: honorary doctorate from Union College, Schenectady; president of the American Institute of Electrical Engineers

(1901), award of a gold medal by the Franklin Institute, member of the American Philosophical Society among others.

Development of modern power systems

"At the center of Steinmetz's life, was his approach to engineering specifically to the complex problems of generating, transmitting and using electric power. By the 1890s, it was already evident to many that the advantages of a.c. over d.c. were compelling in many circumstances, especially when transmission over long distance was involved. But certain specific problems in devising appropriate devices for a.c. and the greater complexity of calculating the behaviour of a.c. circuits stood in the way of its wider application" *(Science,* Vol. 257, 31 July 1992).

The development of modern power systems using alternating current requires the explication of mathematical and electrical concepts that were unfamiliar to engineers themselves at that time. One of his contemporaries summed up ***Steinmetz's contribution*** figuratively, ***"The generating of electricity from the square root of minus one ($\sqrt{-1}$)".***

Nikola Tesla (1856-1943)

Who ushered in the power age

When the U.S. Postal Service decided in 1982 to honour American electrical inventors in its stamps, it chose to depict specific objects they had invented. For Charles Steinmetz (1865-1923), the perplexed stamp designer simply put a graph of a sine wave next to his picture. And with this neat little symbol, they captured the essence of his contributions. ***In the case of Nikola Tesla,*** it was straightforward — ***an induction motor that appeared on the stamp.*** He was the ***pathfinder of the power age.***

Passion for mathematics and engineering

Tesla was born on July 10, 1856, in a mountain village Smiljan, Croatia (till recently in Yugoslavia). His father belonged to the clergy, which profession he intended his son to follow. But the young boy showed an unusual capability in mathematics and science. On graduation from the Gymnasium in Karlovac, he persuaded his father to let him go for higher studies and join the polytechnic at Graz, Austria.

Discovery of the a.c. motor

As a student, he observed in classroom demonstrations that a dynamo run as a motor sparked badly between the commutator and brushes. Tesla suggested ***the possibility of a motor without a commutator,*** an idea that his professor did not endorse.

In 1879, Tesla moved to Prague to enroll at the University but left without taking a degree, when his father died. He then took up a number of jobs to earn his livelihood; in 1881, he worked for the new telephone company at Budapest. ***Here he revived the thought of the principle of rotating magnetic field and discovered the a.c. motor.***

Tesla describes how ***the discovery was instantaneous and intuitive:*** "While walking in a park with a friend Szigety, I was moved to recite a passage from Goethe's Faust, when the idea came like a lightning flash. In an instant I saw it all, and drew with a stick on the sand the diagrams which were illustrated in my fundamental patents of May, 1888; and which my friend understood perfectly." (*Dictionary of Scientific Biography,* Vol. XIII, Ch. Scribner's Sons, New York)

Machinery for the power industry

In 1882, Tesla worked in Paris and Strassbourg. He built (1883) a prototype of his motor and was delighted to see for the first time, rotation effect by a.c. without a commutator. In 1884, he landed in the U.S. to promote his new a.c. motor, "with a book of poetry, four cents in his pocket and an introduction letter to Thomas Edison (1847-1931)". The latter received him cordially; though totally committed to direct current, he gave Tesla a job.

Tesla left Edison in 1885 to work for a company developing and promoting an industrial arc lamp. On the commencement of production, the company quietly dispensed with his services; for some time, he lived precariously, earning by day labour. Within two years, he was back in business and had set up his own laboratory for the development of the a.c. motor.

By 1888, he took patents for a.c. machinery — the dynamo, induction motor and transformer. George Westinghouse (1846-1914) bought these patents and brought to the market place Tesla's polyphase system. Tesla's system provided the basis for the whole modern power industry. The combination of the induction motor and polyphase system provided ***the first practical means for generating a large quantity of electricity at one place and transmitting it economically over long distances.*** So the large-scale harnessing of the Niagara Falls was made possible (1895-1903).

In 1889, Tesla became an American citizen. In 1891, he gave lectures on his invention — the Tesla coil and air-core transformer and repeated them in Europe, winning great acclaim. ***He gave exhibitions in his lab***, in which he lit lamps without wires by ***allowing electricity to flow through his body, to allay fears about a.c.***

National hero

Tesla was indeed a scientific prophet. In a demonstration that he gave in 1893 at St. Louis,— two years before Marconi's experiments — ***he predicted wireless communication.*** He developed torpedoes and ships guided by radio navigation. He forecast in 1917, the radar.

Tesla enjoyed a measure of fame; but ***he made little money from his inventions*** and became increasingly poor during his last years. He was a complete recluse, living in hotel rooms with only pigeons for company. He died in New York on January 7, 1943.

His papers and notes are now in the Nikola Tesla Museum in Belgrade, where he is revered as a national hero. In 1956, the ***unit of magnetic flux density was named 'tesla' in his honour.***

GEOFFERY INGRAM TAYLOR (1886-1975)

A CLASS ACADEMICIAN

He tended to take up a problem that were entirely new, and chose to work alone. Instead of mastering tensor notation, electronics, or numerical computations, G.I. Taylor chose to do things his own day, and did them better than anybody else.

From physics to mechanics

Taylor was born on March 7, 1886 in London; his associates and friends simply refer to him as 'G.I.': he was the grandson of the self-taught mathematician George Boole (1815-1864) who invented 'Boolean Algebra'. ***He inherited the love for mathematics from his mother.***

Taylor read mathematics and physics at the Cambridge University and started his career under J.J. Thomson (1856-1940). But he gave up pure physics and following his natural bent, changed his field to mechanics. He accepted a research position in dynamic meteorology, although he had no knowledge of meteorology!

As a meteorologist

In 1913, at the age of 27, he was invited to serve as a meteorologist on a British ship that sailed to Newfoundland to investigate the sinking of the Titanic. He took the opportunity to make measurements of velocity, temperature and humidity profiles up to 2000 metres by flying kites and releasing balloons from the ship. These were the very first measurements on the turbulent transfer of momentum and heat in the fractional layer of the atmosphere. ***This activity started his lifelong interest in turbulent flows.***

During World War I, the British Air Force commissioned him as a meteorologist. He learned to fly and became interested in aeronautics. He made the first measurement of the pressure distribution over a wing in full-scale flight. Involvement in aeronautics led him to an analysis of the stress distribution in propeller shafts. This work finally resulted in a ***fundamental advance in solid mechanics, the 'Taylor dislocation theory'.***

Research career

Taylor had an extraordinarily long and productive research career (1909-1972). The amount and versatility of his work can be illustrated by the size and range of his *Collected Works* published in 1954:

- Volume I: Mechanics of Solids (41 papers, 593 pages)
- Volume II: Meteorology, Oceanography and Turbulent flow (45 papers, 515 pages)
- Volume III: Aerodynamics and the Mechanics of Projectiles and Explosions (58 papers, 559 pages) and
- Volume IV: Miscellaneous Papers on Mechanics of Fluids'(49 papers, 579 pages).

Perhaps G.I. Taylor is best known for his work on turbulence.

"Taylor combined a remarkable capacity for analytical thought with physical insight by which he knew 'how things worked'. ***He loved to conduct simple experiments, not to gather data to understand a phenomenon,*** but to demonstrate his theoretical calculations; ***in most cases, he already knew what the experiment would show.***"

Taylor was a thoroughly lovable man who did not suffer from the maladjustments and self-concern that many of today's institutional scientists seem to suffer (because of pressure!), and this allowed his creative energy to be used to the fullest extent.

Original style of pursuing research

"He thought of himself as an amateur, and worked for pleasure alone. He did not take up a regular faculty position at Cambridge, had no teaching responsibilities, and did not visit another institution to pursue his research. He never had a secretary or applied for research grants; ***the only facility he needed was a one-room laboratory and one technical assistant.*** He did not 'keep up with the literature', tended to take up ***problems that were entirely new, and chose to work alone.*** Instead of mastering tensor notation, electronics, or numerical computations, G.I. Taylor chose to do things his own way, and did them better than anybody else". (*Biographical memoirs of the Fellows of the Royal Society,* Batchelor G.K, Vol. 22, 1976)

Honours

Taylor received many honours during his life — FRS (1919), Copley Medal, Knighthood (1944), O.M (1969), for his many contributions to aerodynamics, hydrodynamics and the structure of metals.

His most important contribution in 1966 — made at the age of 80 — is the realisation that ***the idealisation of perfect conductors or perfect dielectrics is misleading***. There is always some residual free charge present on the interfaces between different fluids. He died in his 90th year (June 22, 1975) at Cambridge, England.

(***Founders of Modern Fluid Dynamics*** in *Fluid Mechanics*, Kundu and Cohen, Allied Publishers, 2002.)

JAMES WATT (1736-1819)

WHO PROPELLED THE INDUSTRIAL REVOLUTION

Though the invention of the steam engine, the king of machines, belongs to our own epoch, the idea was born about 2000 years ago, in Alexandria. From the state of theory; it was taken in hand by practical mechanics. One of the benefits of Watt's invention was the establishment of textile mills.

Educated at home

James Watt was born on January 19, 1736, at Greenock, Scotland; he was the son of a shipwright and maker of nautical instruments. Due to his fragile health, his mother taught him at home. He suffered from serious attacks of migraine and at school; both his classmates and his teachers took a poor view of his ability.

The source for a more consequential part of his education was his father's workshops where, with his own tools, bench and forge, he skillfully made models of the ship's gear, which he found lying around. ***The workmen acknowledged the boy's promise as a master craftsman*** in the words, "Jamie has gotten a fortune at his fingers' ends".

In 1755, Watt went to London where he spent a year as an apprentice to a mathematical instrument maker, producing navigational and surveying instruments. A year later he returned to Scotland.

It was only through the influence of friends on the faculty of the University of Glasgow that he was able in 1757 to overcome the opposition of tradesmen to establish himself in Glasgow in this trade. He was appointed instrument-maker to the Glasgow University and was allowed to work in a room within its precincts.

About this time, Watt became a friend of the scientist Joseph Black (1728-1799) who gave the concept of latent heat. ***Their friendship was influential in the early development of the steam engine.*** In 1763, Watt was asked by the university to repair a model of an atmospheric engine, a type of steam engine invented around 1700 by Thomas Newcomen (1663-1729) for pumping water out of mines. The engine was very inefficient, extravagant in fuel.

Improvements to the Newcomen engine

After a few experiments, Watt recognised that ***the fault with the model rested not so much on its construction but on its design.*** For two years, he could think of no means of improvement, until a solution flashed in his mind during a sunday afternoon walk.

He proposed that condensation must be effected in a chamber distinct from the cylinder but not connected to it. This is hailed as ***the greatest invention in the whole history of the steam engine***. It took a decade to solve all the associated mechanical problems. Black lent him money and introduced him to John Roebuck (1718-1794) who became his partner in 1765. However, ***Watt had to earn his livelihood as surveyor for canal construction*** and this employment left him little time for developing his invention.

However, he managed to patent his engine on January 5, 1769. By 1773, Roebuck's financial difficulties brought Watt's work to a standstill. Matthew Boulton (1728-1809) who was running the famous Soho Foundry near Birmingham, bought out Watt's patent and began in 1776, to manufacture the engine.

During the next five years until 1781, Watt installed and supervised numerous engines for pumping water in colliery, copper and tin mines, the managers of which wanted to reduce the costs. So, he was obliged to endure keen bargaining in order to obtain adequate royalties.

Further modifications

Boulton, foreseeing a keen market in the cotton mills, urged Watt to invent a rotary motion for the steam engine, to replace the reciprocating action of the original. He incorporated further modifications for a more efficient utilisation of steam: the use of a double-acting piston, rotary motion, parallel motion that guided the piston rod in a perpendicular motion and the centrifugal governor for automatic control of the speed.

The additional invention of a pressure gauge to indicate the steam pressure and the degree of vacuum within the cylinder completed (1790) the Watt engine. Watt became a wealthy man, having received £76,000 in royalties on his patents.

During the 1790s, he had to fight a litigation through which he preserved his separate condenser patent. In 1800, he retired from the business, but continued to work in the garret of his house, till his death on August 25, 1819.

As the steam engines replaced animal power, ***Watt naturally evolved a rational method based on measurement in terms of the number of horses the engine replaced.***

He found the value of one horsepower to be equal to 33,000 pounds raised through one foot each minute; the charge of erecting the steam engine was accordingly based upon its 'horse power' *(The McGraw-Hill Encyclopedia of World Biography,* Vol. 11).

Other inventions

He invented a sculpting machine with which he reproduced original busts for his friends.

There was always a good deal of correspondence to be copied in connection with his business. So, Watt devised a perspective machine for copying letters and drawings, which was displaced by a photocopier and typewriter.

Watt displayed considerable interest in chemistry: he was in contact with the foremost chemists of his time (Black, Priestly and Berthollet). He contributed to the commercial application in Britain of ***the process of bleaching textiles with chlorine.*** He evolved a process, based on the theory given by Black (1728-1799), for producing alkali — using common salt and lime as ingredients. These contributions can indeed be called the chemical equivalents of his mechanical inventions.

Honours

Watt's achievements were amply recognised in his lifetime: honorary doctorate of the University of Glasgow (1806), Fellow of the Royal Society of Edinburgh and London (1785) and Member of the French Academy of Sciences (1814). Watt was elevated to the rank of Baronet, which he declined.

Watt left the most valuable legacy for his country, namely useful machine power for the development and proliferation of industry. ***His name is enshrined in the unit of power — a watt*** is one joule per second, and one horsepower is equivalent to 746 watts.

George Stephenson (1781-1848)

Who invented the steam locomotive

Following the invention of the steam engine by James Watt (1736-1819), George Stephenson pioneered the building of the first railway in England and of the steam locomotive; this marks a milestone in the history of surface transportation.

Early life

George Stephenson was born on June 9, 1781 in a village near Newcastle-upon-Tyne; he was the son of a fireman. He had ***no formal schooling;*** as a boy, he had to look after cattle. ***When he was 14, he became an assistant to his father and worked as colliery engineman.*** But he had great powers of observation, keen mechanical intuition and remarkable foresight. He learned to read and write at the age of 17, by attending evening classes.

Stephenson soon married and in order to supplement his income, he learned miscellaneous jobs — to repair shoes, fix clocks and cut clothes. Later when he sent his son Robert to school, he tried to make up for the lack of formal education — every night when the boy came home, father and son went over the homework together.

In 1808, when the prospects of employment seemed bleak, he seriously thought of emigrating to America. Fortunately, he obtained a contract for improving the performance of a Newcomen engine at Killingworth colliery. He took the engine to pieces in order to understand its construction. The reward came with his success: he was appointed in 1812 as engine-wright (chief mechanic) to the colliery.

Stephenson's ***first invention was a safety lamp for miners*** (1815); which was developed at about the same time by Humphrey Davy (1778-1829). He tested it in the gas-infected tunnels. This led to a controversy as to who was the real inventor, but before long, it was acknowledged that both men had reached the same goal by different approaches.

The Blucher

In 1813, Stephenson visited a colliery in Yorkshire to examine a 'boiler on wheels', constructed by John Blekinsop (1783-1831) for hauling coal wagons. It employed a ratchet wheel running on a cogged third rail, an arrangement that caused frequent breakdowns. He carried out modifications and produced in 1814 a smooth-wheeled locomotive, called 'the Blucher', which drew 30 tonnes of coal at four miles per hour (mph). Not satisfied with the locomotive power he could achieve, he introduced the 'steam blast' by which steam was redirected up the chimney, pulling air after it and increasing the draft through the fire. The new design made the locomotive truly practical.

Subsequently, Stephenson started to make experiments with various gradients. He discovered that a slope of 1 in 200, common enough on roads, reduced the haulage power of a locomotive by 50%. Thus, he concluded that railway gradients should always be as low as possible. So cuttings, tunnels and embankments were necessary.

Architect of rail transportation

In 1821, Stephenson heard of a project for a railroad (employing draft horses), to be built from Stockton to Darlington to facilitate exploitation of a rich coal mine. He convinced Edward Pease, the promoter, that steam locomotion was better than the proposed horse traction. Accordingly, Pease commissioned him to build a steam locomotive for the line. ***The gauge for the railway was assessed by Stephenson at 4 feet, a historic decision***; this has remained the standard gauge for railways throughout the world.

On September 27, 1825, rail transportation was born with the opening of the Stockton–Darlington line. Stephenson's engine 'Locomotion' took a party of 450 passengers at a top speed of 15 mph.

An innovation

Stephenson was next called to build a railway connecting Manchester to Liverpool, as the trade exceeded the capacity of the existing canals. For an efficient locomotive, he suggested ***increasing the heating surface of the boilers by tubes*** passing through the cylindrical barrel. The new engine 'Rocket' was built under the direct supervision of his son and weighed 4 tons, half the weight of his earlier model 'Locomotion'. It reached a speed of 14 mph.

In the battle between the supporters of the locomotive and the atmospheric railway system evolved by Brunel (1806-1859), Stephenson's supporters in Parliament carried the day (1845); ***this ended the last attempt to challenge the steam locomotive.***

Railroad construction spread rapidly in Britain, Europe and North America. Stephenson continued as ***the architect of the revolutionary transportation system,*** solving problems of railroad construction, bridge design, locomotive and rolling stock manufacture.

Till the end of his life (August 12, 1848), he worked as a consulting engineer to railway companies. He was the first president (1847-48) of the Institution of Mechanical Engineers, London.

Sadi Carnot (1796-1832)

Who completed the engine cycle

In 1847, Joule proved that work could be turned into heat. This constitutes the first law of thermodynamics. Only 10% of the heat in the fuel could be turned, by the early engines, into useful work. Scientists asked why the reverse process was proving to be so inefficient. The answer was provided by the fundamental work of two of Joule's eminent contemporaries, Sadi Carnot and Rudolf Clausius.

Turbulent period in boyhood

Sadi Carnot was born in Paris on June 1, 1796. His father served as a minister of the interior under Napoleon. The turbulent worlds of French politics and science were an integral part of the environment which shaped his youth. Carnot entered the elite Ecole Polytechnique, Paris, at the age of 16.

He studied there from 1812 to 1814, and ***his teachers were well-known men of science*** — Poisson (1781-1840), Gay-Laussac (1778-1850), Ampere (1775-1836) and Arago (1786-1853). He acquired his taste for mathematics, besides a solid training in physics, natural sciences and languages.

In October 1814, Carnot was sent to the military school at Metz. On completing the two-year course in military engineering, he began serving in the engineering regiment. For the next two years, he was sent from one garrison to another to do routine engineering work. The tedium of work in military garrisons thwarted his intellectual development.

In 1819, he passed a competitive examination for admission to the corps of army staff officers. He immediately obtained a permanent leave of absence, took up residence in Paris and began a wide range of study and research, which was interrupted in 1821, by a few weeks, when he visited his father in exile in Germany.

Treatment of basic questions

Carnot's interest turned more and more towards basic questions, which led to his masterpiece *Reflexions.* In 1828, he resigned from the army, settled in Paris and focussed his attention on engine design and theory of heat — the physical properties of gases and vapours and the relationship between temperature and pressure.

Carnot's earliest work is a 21-page paper entitled *Recherche.* It was written probably in 1823, being the first mathematical attempt to find an expression for the work produced by one kilogram of steam as a function of temperature alone. ***The essay is distinguished for the clear analysis of concepts and use of both an adiabatic working stage and an isothermal stage.*** The manuscript, however, ***remained unknown to the scientific world until 1843.*** The other major work *Reflexions* that Carnot published appeared in 1824.

Carnot recognised that previous work on steam engines had made use of an abstract three-stage cycle, namely isothermal and adiabatic expansions and isothermal compression. Carnot saw that the cycle was incomplete and ***pushed ingeniously the abstraction one step further by adding a final adiabatic compression,*** in which work was consumed to heat the steam back to its original temperature.

In June 1832, he contracted scarlet fever and brain fever. He died on August 24, 1832 in the prime of his youth.

Thermodynamic concepts

Thus Carnot introduced two important thermodynamic concepts — ***completeness and reversibility.*** At the end of each cycle, the engine and the working substance returned to their original conditions. This complete cycle not only provided an unambiguous definition of the input and output of the engine, but also rendered superfluous the examination of each stage of the cycle. With each cycle, the engine transferred a certain quantity of heat from the high-temperature condenser, thereby producing work.

Thus the 'Carnot engine' was born, consisting of a cylinder and piston, a working substance that he assumed to be a perfect gas and two heat reservoirs kept at different temperatures.

Carnot enunciated his result in a fundamental theorem known by his name, "the motive power of heat is independent of the agents employed to realise it, its quantity is fixed solely by the temperatures of bodies".

In 1834, Clapeyron published an analytical reformulation of the *Reflexions.* ***Still it was not until 1843,*** when the German translation of Clapeyron's review paper appeared in the *Annalen der Physik und Chemie,* ***that the world of science noticed Carnot's fundamental work.*** By then, Carnot had been dead for 11 years!

Extension of Carnot's work

William Thomson (1824-1908) in 1848 began working on special aspects of Clapeyron's paper and on *Reflexions* and ***published a series of papers that confirmed and extended Carnot's result.*** In 1830, Clausius (1822-88), another physicist, showed that Carnot's result, which assumed no loss of heat, needed modification. In the Carnot engine, a certain quantity of heat is destroyed, another quantity is transferred to the cold body, and both quantities are related to the work done.

With these additions, adopted in 1851 by Thomson, ***Carnot's theory marked the beginning of the science of thermodynamics.***

Robert Fulton (1765-1815)

Who pioneered steam navigation

Robert Fulton has often been assumed to be the inventor of the steamboat, as James Watt (1736-1819) is generally believed to the inventor of the steam engine. Recorded history of science shows that this notion is incorrect. These two inventions are the results of the inventive powers of the human race, not the outcome of the creative genius of any single individual.

It is an undeniable fact that Watt added many innovations to the steam engine, as it came from the hands of Newcomen (1663-1729). As an inventor, Fulton did comparatively little in the realm of intellect; he did nothing to modify the engine or improve the steamboat.

Fulton simply took the products of other innovators and set them to work in combination, which resulted in a number of useful applications. It is this achievement, which made ***Fulton a celebrity in the engineering world. He made possible moving ships by steam on a commercial scale.***

Taught by his mother

Fulton was born on November 14, 1765 in Lancaster, Pennsylvania State. His father worked at farming, in which he was not successful. He died in 1768, when Robert was three years old. He was taught at home by his mother, a woman of superior attainments, till the age of eight, when he went to a Quaker school. Fulton did not distinguish himself at school. By the age of 10, he showed promise as an artist: be was employed by local gunsmiths to make drawings for their work.

Painting as a career

In 1782, he moved to Philadelphia and spent four years making portraits and doing miniatures. He bought a farm near the city for his widowed mother.

In 1786, Fulton proceeded to London to study painting with his family friend Benjamin West, one of the leading painters then living in England. He came at a time when England was in the midst of the industrial revolution. So Fulton was spontaneously fascinated by what he saw — canals, mines, bridges and factories. So he gave up in 1793 painting as a profession and ***turned his attention to engineering projects — studies on the English canal system, steam navigation and marine propulsion.***

Fulton was drawn to the first problem by perusing a paper on a canal to join the Bristol and English Channels (1793) projected by Lord Stanhope. The difference of levels was upwards of 500 feet and water was scarce.

Fulton wrote to His Lordship, enclosing a sketch of a scheme for doing away with the necessity for locks and minimising the loss of water. This scheme was to consist of a prepondering cistern of water to draw the canal boat up an incline from one level to another. The inclined plane was not new: indeed, it was in use for canals in China a thousand years ago. In its simplest form, it is seen at many of the locks on the upper Thames.

Made naviagation history

Next, Fulton considered ***the problem of moving boats by means of steam power,*** which had far-reaching influence on his own career and on the course of navigation history.

In Fulton's own words, "the repeated failure of efforts by men of science to move boats by the power of steam engines gave an impression to the public mind both in Europe and America, that it was impracticable to make a useful steam boat, and under this belief ***those who attempted it were considered as visionaries or mad men".***

Fulton tried in 1801 to interest the French Government in his experiments. It is doubtful whether Fulton would have done anything in steam navigation, had it not been for the arrival in 1802 of Robert Livingston as U.S. Minister to the French Government. Being deeply interested in this field and as a result of his experiments, he agreed to support Fulton's idea of building a steamboat.

In 1805, Fulton contacted the firm Boulton and Watt for the supply of a steam engine for boat propulsion. But it took another three years to obtain permission for exporting the engine, which the British government had banned. Fulton was able to sail for America only in 1806 with the engine assembled from many parts.

The double-acting rotative engine developed by James Watt was selected for the first time the horsepower obtainable from an engine and its boiler, (per pound weight and per cubic feet occupied) was reduced to within the limits of displacement of a boat that the engine could propel at a reasonable speed. Without these ingenious modifications, the engine would have sunk the boat by its own weight!

The engine was put together in New York and set aboard a locally built vessel. Although Livingston had been an advocate of a kind of jet propulsion for steamboats (that is, a jet of water forced out the back of the boat under high pressure), the two now settled on paddle wheels. Under way, it averaged five miles per hour. ***On August 17, 1807, the 'Clermont' began its first successful voyage up the Hudson River to Albany.*** After the voyage of the 'Clermont', steamboats appeared up and down the Atlantic Coast.

Before his death (February, 24, 1815), Fulton had built a large boat works in New Jersey and directed the construction of 17 steamboats and a torpedo boat.

Canal navigation in America

The first important steps to introduce canal navigation could be reailsed only in America where the extensive waterways but ***lack of highways suggested that transport by water was easier and cheaper than by land.*** The impetus might have been due to the removal of trade restrictions after the War of Independence (1175-1783).

Fulton patented in England a double-inclined plane for hauling canal boats over difficult terrain. He built a mechanical dredge to speed up the construction of canals. He published amidst his various activities his monograph 'A treatise on the improvement of canal navigation' (1796).

For the next 10 years, Fulton devoted his time to the development of underwater warfare, through the invention of the submarine and the self-moving torpedo. He tried to interest the French Government and later the British Government in his experiments but was unsuccessful.

In this venture, ***his motto*** had been, ***"The liberty of the seas will be happiness of the earth".*** Naval warfare, once rendered impossible, the freedom of the seas was assured, which he felt would make safe the liberty and prosperity of his native country.

The 'diving-boat', as Fulton called it, seems to have been remarkably successful judging it by even modern standards.

What contributed to Fulton's success

To analyse the causes for Fulton's success where at least a dozen American inventors failed: in Livingston, he had a trustworthy and politically powerful person who was able to obtain a lucrative monopoly on the steam navigation of the state's waters.

Previous inventors such as John Fitch had to build their own engines, as the art of constructing steam engines was at that time totally unknown in America. But Fulton could start his work with an engine supplied by the world's leading builders.

Besides, he was able to employ mechanics who had over 20 years' experience with the steam engine. It was Fulton's genius that could combine these elements into a successful steamboat enterprise. (*The McGraw-Hill Encylopedia of World Biography*)

Charles Babbage (1792-1871)

Architect of modern computers

"The earliest idea that I can trace in my own mind of calculating arithmetic tables by machinery arose in this manner. One evening, I was sitting in the rooms of the Analytical Society at Cambridge, in a kind of dreamy mood, with a table of logarithms open before me. Another member, coming into the room and seeing me half asleep, called out 'What are you dreaming about, Babbage?' ***I am thinking that all these tables might be calculated by machine***", so wrote Charles Babbage. (*Life of a Philosopher,* Dover, 1968)

Prior to Word War II, the need for computational power was seen only in astronomy and physics. From the beginning of the First Difference Engine he conceived, and with the ideas that followed in quick succession culminating in the Analytical Engine, Babbage established his position as the architect of modern computers.

His career

Charles Babbage was born in Teignmonth, England, on December 26, 1792. When he was 20, he helped found the Analytical Society, whose purpose was to introduce developments from Europe into English mathematics. He settled down with his wife in London in 1820. Babbage lived mostly on a private income provided, though reluctantly, by his banker-father. He also served as Lucasian Professor of Mathematics at Cambridge University during 1828-1839, though he did not deliver a single lecture in those 11 years of office.

Building the difference engine

He was offered a post by a life insurance company because of his mathematical abilities. (He complied the first reliable actuarial tables.) But he turned it down and clung to his aspirations for building his difference engine.

In his pursuit of the Analytical Engine, Ada Byron, later Countess of Lovelace, was his close associate and devoted herself to being his publicist. To strengthen her knowledge, she attended a series of lectures on the engines and strove to educate herself with the help of tutors in mathematics, so that she could understand the engines completely. ***She left behind the most complete account of the Analytical Engine available today.*** Among Babbage's eminent friends were Michael Faraday (1791-1867), William Herschel (1738-1822), Charles Darwin (1809-1882) and George Stephenson (1781-1848).

Other contributions

Babbage made notable contributions in other areas as well:

- the cowcatcher for trains,
- improvement of the safety of Brunel Railways,
- establishment of the modern postal system in England,
- the flashing lighthouse,
- the speedometer and
- techniques for operational research (applying objective criteria to decision making).

Babbage took active interest in public life and was ***a leader of the Science Reform movement.*** He published two books — *On the Economy of Machinery and Manufactures* and *Reflections on the Decline of Science and its Causes.* To reform the society, he wanted to begin with schools and universities. His observation rivets our attention. "A young man passes from our public schools to the universities, ignorant almost of the elements of every branch of useful knowledge".

Babbage wrote on a wide range of subjects — philosophy, geology, archaeology, optics, lighthouses, armaments, postal services and chess. By the time he died on October 18, 1871, he was a disillusioned man, blaming the Government for its lack of appreciation of his ideas.

The dreamer

The so-called Analytical Engine is actually a mechanical device designed by Babbage during the 1830s, to combine basic arithmetic operations with decisions based on its own computations. Babbage's design was forgotten until his writings were discovered a century later, in 1937. ***Had he built it, it would have been the size of a locomotive and would have required a steam engine to power it!***

But Babbage could not accomplish the actual realisation of his grandiose project. The reasons for his failure are attributed to his prickly personality, paucity of funds, personal vendettas and a dispute with his engineer. Babbage's failure to build any of his engines received as much publicity as the originality of the inventions, in *the bi-centennial celebration of his birth, 1992.*

The ***Difference Engine Number 2*** was assembled from 4000 parts. It weighed three tonnes, was 11 feet long, seven feet high and 18 inches deep and was ***built for the first time 150 years after it was designed. It is on display*** at the Science Museum in London ***as a monument to Charles Babbage, the genius and dreamer.***

Irving Langmuir (1881-1957)

Penchant for the unusual

'A saving of a million dollars a night!'—Irving Langmuir made this a reality in the electricity bills of Americans through his discovery that the life of incandescent lamps could be extended by filling them with small quantities of neon or argon gases. This gave General Electric enough financial returns to acknowledge his discovery in an unusual way. Langmuir was allowed to pursue his own research interests for the rest of his career.

Why he chose metallurgy

Irving Langmuir was born on January 31, 1881 and received his education in the Columbia School of Mines. That he had an incisive mind was apparent from the unusual method he applied in selecting his course of studies. ***He chose metallurgical engineering rather than pure science*** not because it was strong in chemistry, but as it contained 'more physics than a straight-forward chemistry course, and more mathematics than a straightforward physics course'. ***This decision as a young student had much to do with what he accomplished in his later research career*** — introducing more physics and mathematics into chemistry.

Higher education in Germany

Langmuir joined the University of Gottingen in 1903 for his doctorate under Walther Nernst (1864-1941). He could appreciate the spare time made free for private studies in German Universities, as he remarked in a letter to his mother: "The life here is totally different from that in America in that it is always interesting. Instead of working all day in college, as I did at home and only studying during the evenings, here I have an hour every morning for study and all day on wednesday and saturday". His research dealt with the behaviour of gases in the presence of a heated filament.

The work that changed light-bulb technology

Langmuir was able to continue work in the newly established General Electric Research laboratory, where he started his career in 1909 and remained until his retirement in 1950. The incandescent lamps were evacuated and their life was short. By investigating how to increase their lifetime, he changed light bulb technology completely. It is believed that ***Langmuir's example may well have played an important part in General Electric's acceptance of basic research*** as an integral part of company philosophy.

Growing out of his work in extending the lifetime of electric bulbs, other momentous discoveries followed. He developed the first oxide-coated emitter, which made possible economical manufacture of high power vacuum tubes. He thus contributed largely to the development of radio broadcasting by the construction of high power radio transmitting tubes.

Other discoveries

Langmuir's contributions covered several areas:

- surface properties of liquids,
- catalytic effects of adsorption of films on solids,
- Child–Langmuir law for electron emission,
- Langmuir trough for the study of monomolecular films on the surface of liquids and
- study of ionised gases.

He also ***coined the word 'plasma,'*** now a popular term for ionised gases.

Langmuir received the ***Nobel Prize for Chemistry in 1932*** — the ***first industrial research scientist*** bestowed with this honour. He died in 1957. A mountain in Alaska had been named after him, as a tribute to his lifelong interest in the environment.

Guglielmo Marconi (1874-1937)

Who invented the radio

> ***Next to the industrial revolution*** stands the technological revolution of the twentieth century, in which the ***radio occupies undoubtedly the first place.*** This invention brought people close, brought entertainment to the doorsteps of homes and round the clock, made it possible to communicate worldwide at affordable cost and hence contributed to the happiness and welfare of human society. The invention of the radio by Marconi is thus at ***the center of the developments in electronic media*** that we witness today.

Flair for practical work

Marconi was born at Bologna, Italy, on April 25, 1874, the son of a wealthy Italian and an Irish mother. He did not have any formal education, other than the technical school he attended for a short time. ***He was educated at home by private tutors and his mother.*** Even as a boy he was fond of physics, especially electricity; and showed a flair for practical work, namely building his own equipment and experimenting at home.

How he started working on Hertzian waves

It was by a chance that Marconi saw the obituary note on Heinrich Hertz (1857-1894), dealing with the experiments verifying Maxwell's theory. Hertz had demonstrated the very generation of electromagnetic waves. ***He attended the memorial lecture on Hertz by Oliver Lodge (1851-1940), which spurred***

him to pursue work on Hertzian waves. Leaving aside the fundamental nature of electromagnetic waves, ***he directed his attention to a specific aspect, the distance over which they could be detected,*** and used in telegraph transmission. So he began in 1895 laboratory experiments, encouraged by his mother: these were conducted from his attic at his father's villa at Pontecchio, where he succeeded in sending wireless signals over a distance of one and a half miles.

Birth of wireless telegraphy

Marconi demonstrated in September 1895 his telegraph system to the Government of his own country. Finding that they did not evince any interest, he took his apparatus in 1896 to England and gave a successful demonstration to the British Post Office in London, on Salisbury Palm and across the Bristol channel. Later that year, he was granted the patent for the first practical system of wireless telegraphy. In 1899, he established wireless communication between France and England across the English Channel. Marconi's Wireless Telegraph Company was registered in 1900 in England, the same year he took out his famous 'four sevens' patent No. 7777 for 'tuned telegraphy'.

On an historic day in December 1901, Marconi employed his system for transmitting the first wireless signals between Poldhu in Cornwall (England) and St. John's in Newfoundland (Canada), a distance of 2100 miles. ***By this single achievement, he became instantly a celebrity in the world.*** Between 1902 and 1912, Marconi patented several new inventions: in particular, the magnetic detector (1902), an improvement over the coherer that Oliver Lodge devised, the horizontal directional aerial (1905) and a 'timed-spark' system for generating continuous waves (1912).

In 1902, Marconi equipped the S.S. Philadelphia and sailed to America. While sailing, he received signals from Poldhu and observed that the strength of the received signal increased at night and followed the curvature of the earth.

In 1914, he was commissioned in the Italian army and was awarded the Italian military medal in 1919 in recognition of his war service. During this period, he returned to his investigation of short waves. After an intensive series of trials (1923) between the Poldhu station and his yacht cruising in the Atlantic and Mediterranean, he established the beam system for long distance communication. ***The first beam station linking England and Canada was opened in 1926.***

Opening of the first radio telephone

Marconi began in 1931, research into the propagation characteristics of still shorter waves, which resulted in the opening of the first microwave radiotelephone (1932) between Vatican and the Pope's summer residence. He extended the use of radio communication for ship navigation (1934) and gave in 1935 a practical demonstration in Italy of the principles of radar. He had forecast earlier (in 1922) the birth of radar, in a lecture before the Institute of Radio Engineers, New York.

Honours

Marconi received numerous international honours and awards, chief among them being the ***Nobel Prize for Physics*** (1909), which he shared with professor Carl Braun (1850-1918). He was decorated with the hereditary title to Marchese (1929). Marconi was a shrewd businessman, promoted business enterprises and amassed ***a tiny fortune from his inventions.***

Marconi was proud of his nationality and supported the Fascists. Before the rise of Fascism in Italy, he died in Rome on July 20, 1937. ***All radio stations in the world observed a two-minute pause to honour him.*** A memorial now stands on the cliff at Poldhu (England), from where Marconi made his trans-atlantic transmission.

Edward Appleton (1892-1965)

Pioneer of radio communication

He showed that radio waves of sufficiently short wavelength can penetrate the lower region of the ionosphere but are reflected by an upper region. This discovery made possible reliable long-range radio communication.

College career

Edward Appleton was born on September 6, 1982 at Bradford, England. He showed exceptional promise in school and won a scholarship to enter the Cambridge University. At St. John's College he attended the lectures of J.J. Thomson (1856-1940) and Ernest Rutherford (1871-1937). He graduated in 1913 with a first class honours degree in physics. He served as a signals officer, during World War I, when he got interested in the radio.

Academic positions

He returned to Cambridge and worked at the Cavendish Laboratory on vacuum tubes, till he was appointed Wheatstone Professor of Physics at the University of London in 1924, where he served the next 12 years.

In 1936, Appleton moved to Cambridge as professor of natural philosophy. With the outbreak of hostilities in 1939, he entered the government service as secretary, Department of Scientific and Industrial Research and played a major role in war efforts. Appleton became, in 1949, vice-chancellor of the University of Edinburgh where he remained for the rest of his life.

The ionosphere

Appleton had begun in 1924, a series of crucial experiments that proved the existence of the ionosphere. With the collaboration of the British Broadcasting Corporation (BBC) and its transmitter at Bournemouth, he shot radio waves into the atmosphere to see if they were reflected by it and if they returned to the ground. The position of the reflecting layer (known as the E layer), which had been postulated earlier by Heaviside and Kennelly, was identified at a height of about 97 km. The ionosphere was thus the first object detected by radiolocation and led to ***the birth of radar,*** a military invention of greatest importance in World War II.

In 1926, Appleton discovered another layer, which is considered to be of prime technological significance. He showed that radio waves of sufficiently short wavelength can penetrate the lower region of the ionosphere but are reflected by an upper region (known as the F layer), at a height of 240 km above the ground. ***This discovery made possible reliable long-range radio communication.***

Appleton furnished experimental proof that the interference between the ground and sky waves resulted in the ***radio fading*** in and around 100 km from a transmitter, in a manner analogous to optical interference fringes.

Honours

Appleton's scientific achievements were honoured by his election to the Royal Society (1927), a knighthood (1941) and the award of the ***Nobel Prize in Physics (1947)*** for his investigations of the physics of the upper atmosphere, especially the discovery of the Appleton layer. He spent the rest of his life in research work flowing from his own discoveries, maintaining a degree of involvement that was astonishing, in view of the many other responsibilities he carried.

Appleton was an international figure. ***He was instrumental in organising the first International Geophysical Year in 1957,*** a year of maximum sunspot activity. He founded in 1949 the *Journal of Atmospheric Research* (affectionately known as Appleton's Journal) and served as its editor till his death on April 21, 1965.

Patrick Haggerty (1914-1980)

His gift for Christmas '54

On July 1, 1948, Bell Laboratories publicity announced the invention of the transistor. Till the transistorised radio arrived in time for the Christmas market of 1954, this celebrated invention lay commercially neglected. The person who made ***the change-over from the vacuum tube to the transistor*** possible was Patrick Haggerty, the key figure in the innovation.

Education and career

Born in 1914, he received his B.S. and M.S. degrees from the Marquette University, U.S. in 1936 setting up an all-time record for marks in electrical engineering. After serving in the Navy's Bureau of Aeronautics during World War II, he joined the Geophysical Service Inc. in 1945. This eventually became the famous Texas Instruments Inc. (TI). He became successively general manager of the manufacturing division, director and president. He was honorary chairman of TI, till his death in 1980.

Launching the transistor radio

Haggerty could intuitively foresee that the days of the bulky vacuum tube in radio equipment were numbered. ***"We needed soon to find an application which would demand from us large quantities of semiconductor devices*** of adequate quality at prices which were economic for the applications involved.

It was to further this goal that an R&D programme to build an all-transistor pocket radio was launched", so wrote Haggerty prophetically. His thoughts were realised with the manufacture of the Regency transistor radio. It must be remembered that this was at a time when one could buy a vacuum tube radio for $15, while a transistor radio was $49.95.

His innovation

Haggerty and his colleagues spent most of 1951, in pursuit of buying the transistor license from Bell laboratories. ***Their presence at Bell meetings evoked surprise and amusement,*** as the company's scientists wondered if TI could succeed in the business. TI were finally given the license for a payment of $25,000 in the fall of 1951. Six months later, TI produced its first transistor, and by the end of 1953, they succeeded in mass-producing grown-junction germanium transistors. This, at that time, meant turning out batches of a thousand!

Haggerty conceived the ***idea of the transistorised radio*** and it was at his instigation that the programme got underway in the spring of 1954. ***All his efforts to interest major radio manufacturers in the U.S. failed:*** their attitude was wait-and-see.

Production of pocket radio

He followed his thoughts with unusual strength and persistence. In spring of the same year, he saw a magazine advertisement from the Independent Development Engineering Associates (IDEA), an Indianapolis company, and found that they were prepared to produce the radio. He had the prototype built, tested the circuit and gave it to IDEA. Weighing less than 12 ounces and fitting into a shirt pocket, the TR-1 radio was an instant success; in fact, a sensation. It opened up ***the first high volume market for solid-state products.***

It was not long before an agreement was reached for IDEA's Regency division to manufacture the radio with transistors bought from TI, as Haggerty wanted a product out in time for the Christmas market. This was a tall order for a group that as yet had no experience in transistor circuit design, no transistors capable of operating reliably in a radio circuit and of facilities for mass producing such radios. The manufacture of the ***Regency transistor radio heralded the beginning of a new era of electronics technology.***

In later years, TI's strategy would be ***widely cited as a model for technology*** business planning. It was also a strategy that may well have saved a couple of years of the time it would otherwise have taken the U.S. and the rest of the world, to shift from vacuum tubes to semiconductors.

Why not transistors?

One way in which Haggerty propelled the sales of TR-1 was through the arrangement he made with Thomas Watson Jr. around 1955 to buy a few hundred radios and distribute them among IBM's executives. The transistor radio set led to an important licensing agreement with TI that resulted in their becoming a supplier of transistors to IBM for many years.

Thomas Watson Jr. recalled, "Our people had barely learned how to handle electrons in the tube and the transistor invention shocked them". It was something they pulled back from, I used to go up to the lab and say "Why not transistors?" Finally he issued an executive order ***"After June 1, 1958, we will build no more machines with electron tubes".*** That is how transistors found their way to make the IBM computer work.

A technological prophet

Thus Haggerty opened up the consumer sector for high volume solid-state products and spearheaded developments in new electronics technology. In talks and technological papers, ***he articulated his thoughts and predictions concerning Integrated Circuit Technology*** (*Proc. IEEE,* December 1964).

Haggerty observed in 1966, "that the entire cycle of utilisation of semiconductor devices in the U.S. and the world was speeded up at least two years because, the ***then-relatively small company (TI) chose the proper strategy and followed through with successful tactics".***

Formation of IEEE

Haggerty was the driving force for merging of the IRE with AIEE, which led to the formation of the Institution of Electronics and Electrical Engineers (IEEE), the largest professional body in the world. He served as chairman of the Board of Rockfeller University from 1975 to 1980.

In hindsight, Haggerty wrote, "TI and IDEA had an agreement that if we chose, we would stay in the consumer business... had we the funds to go on and develop additional products, we would have entered then in 1954 instead of 1971 and I think the likelihood is very high that we would have been the SONY of consumer electronics". (*IEEE Spectrum,* December 1985)

ARTHUR COTTON (1803-1899)

DEDICATED IRRIGATION ENGINEER

His dedicated work has turned millions of acres of barren lands into flourishing farm lands in Andhra Pradesh. No wonder millions of people worship his statue at the dam site.

Career in the Madras Presidency

Arthur Cotton was born on May 15, 1803 as the tenth son of Henry Calveley Cotton. He joined the Corps of Royal Engineers, when he was only sixteen. His recruitment was without even an examination, as he had a good record during the training period.

Cotton arrived in Madras in September 1821 as a young man of just eighteen and was attached to the office of the chief engineer of the then Madras Presidency. In 1822, he was sent as an assistant to the superintending engineer of the Tank Department in the Southern Division. His responsibility was to repair and maintain the tanks in Coimbatore, Madurai, Tiruchi, Thanjavur and Tirunelveli districts.

He was made Lieutenant in 1824 and was in charge of the military buildings at St. Thomas Mount. He was deputed to serve in the first Burmese War.

In 1826, Cotton came back to Madras as superintending engineer, Central Division of Tank Department, and was also in charge of the Pamban pass work. Later, he was promoted as captain and made in charge of the Cauvery Irrigation and Pamban Pass work. In 1830, he constructed the scouring sluices in the Grand Anicut and in 1832; he planned the Upper and Lower Coleroon anicuts.

Barrages across Cauvery

The old stone anicut (Grand Anicut as he called it) on the sandy bed of the Cauvery and its successful functioning for centuries always amazed him. That was a time when the technology of building a hydraulic structure on a sandy river bed had not developed, he had an innate desire ***to probe the foundations and see for himself how great builders of the second century could have accomplished this engineering feat.***

He designed and constructed the five scouring sluices on the eastern end of the anicut and while excavating for the foundation, found to his surprise that the Grand Anicut had stood all along on foundations laid with large boulders embedded in the layer of stiff clay underlying the sandy bed. The sluices which he introduced in the anicut help in scouring out the silty sand to keep the mouths of the Cauvery and the Vennar clear of sandy mounds for efficient drawal of irrigation waters.

The construction of the Upper Anicut across the Coleroon at Mukkombu (also in Madras Presidency) ensured diversion of adequate irrigation supplies into the Cauvery, while the Lower Anicut downstream on the Coleroon diverts water, which was used for irrigating lakhs of acres of fertile land fed by the Veeranam tank. His superior officer wrote of Cotton's works during this period as those executed with originality and courage, loyalty and conviction.

"The permanent prosperity of Thanjavur is without doubt to be attributed in large measure to that first bold step taken by Cotton in the construction of the Upper Coleroon dam, under circumstances of great difficulty with restricted means against much opposition and with heavy personal responsibilities."

Cotton gave the idea for creation of the Madras harbour by forming a rubble mound breakwater parallel to the coast. He laid the Madras–Red Hills railway line to bring road materials to Madras, incidentally this was the first railway laid in India. Soon after completion of this project his health broke down and he went to Australia.

While in Australia, he married Elizabeth Learmonth on October 29, 1841. This noble lady always stood by her husband, bearing all the pains of living in inhospitable places.

Structure on Godavari

On return from leave, Cotton was posted for a light job of building a church at Waltair. This gave him an opportunity to study the coast and the Godavari delta where he was destined to create his masterpiece in engineering. He made a number of groynes (breakwaters) on the Vizag coast, which proved quite useful.

The Godavari was to be the scene of his most strenuous labours, his keen anxieties and his grand success. This district was affected by famine frequently and Cotton thought it odd that this region should suffer from famine, when a river carrying millions of cusecs of water was flowing through it. He planned the Dowaleshwaram Anicut across it. The river is very wide at the anicut site with three islands occupying more than a third of this length.

Cotton's experience with the Cauvery works and his study of the Grand Anicut founded on sand gave him the courage and conviction that similar works could be built across the Godavari too, though it is much larger with a flood discharge of up to 15 million cusecs. Cotton sent his report in April 1845 to the East India Company in London, which then was sanctioned.

God's mission

This was a mammoth work organised by him and completed in record time. He believed that it was God's mission given to him to execute this job. The anicut work started in 1847, was completed in 1852 and drew great appreciation from the Governor in Council and the Court of Directors of the East India Company in England.

The Dowaleshwaram Anicut served for more than a century diverting supplies for irrigating a million acres and turned the famine-ridden region into fertile districts of the Godavari estuary. It is no wonder that millions of people worship his statue at the dam site. This anicut has now been replaced by a barrage.

In his endeavor, Cotton was not sailing smooth: he had to work against odds. For this work, which had obvious benefits, he had to struggle in convincing the authorities who had the power in their hands, but no vision. Explaining the need for irrigation and drainage, ***he emphasised that irrigation would always mean the complete regulation of water, including draining.***

So there is never a season when there is not excess rain, which needs to be carried off by a system of drains. It is this regulation of water that is needed and which so abundantly repays the cost of works.

His dedicated work has turned millions of acres of barren lands into flourishing farmlands in Andhra Pradesh and he is considered one of the greatest benefactors of India.

The one ambition he held close to his heart was to create large reservoirs, dig a network of navigable canals all over the Indian subcontinent. He strongly believed that it was better to plan a network of navigable canals than long lengths of railways, and vehemently advocated that the large investments in railways denied the native people cheap transport for marketing their produce.

Other activities

Cotton was appointed chief engineer of the Madras Presidency by the end of 1853. Later, he was appointed Commandant of Engineers at Madras without any special functions and was an informal adviser to the Government on engineering subjects. He was the chief authority on all matters connected with irrigation, not only in the Madras Presidency but throughout India.

He prepared a report on an irrigation project in the Province of Orissa, planned dam across the Ganges at Rajmahal (in Uttar Pradesh) and a canal from there for irrigation, navigation and water supply to Calcutta. He also made a report on the Ganges canal, which was under execution by Sir Proby Cautley.

Cotton left India in 1860 and on arrival in England was knighted; he was honoured at a public banquet in London.

Continued interests in India

Even though he retired from his arduous labours in India, his sympathy was so bound up with the interests of the people in India, that he was always considering what more might be and ought to be done. He laboured incessantly both with his pen and by interview with those in power convincing them of the need for creating large storages and connecting canals all over India for extensive irrigation.

Sir Arthur Cotton was a true Christian, with abiding faith in his religion and commitment to serve humanity. During the last 30 years of his life, he warmly supported the temperance cause. Even on his deathbed, he refused to touch any intoxicant. He was cheerful in spite of several calamities in his life and remained active till his death in July 1899.

Bless Cotton Dorai!

There apparently was a Vedic pandit living on the banks of the Godavari who ended his prayers every day calling on the ***Almighty to bless 'Cotton Dorai' with long life and prosperity.*** One day, an English official on circuit who was passing by, stopped on hearing this invocation and wondered how Arthur Cotton figured in the pandit's prayers. If I can take my bath before my daily prayers, it is because of the ***Dowleswaram Anicut he built,*** replied the pandit. But he was only doing his duty in building the dam, responded the officer. It was more than duty; it was dedication and devotion to the people of a country not his own, insisted the pandit. And the official humbly replied, ***"Cotton Dorai, who stands here before you, thanks you for your prayers.*** Will you bless him?"

(*The Hindu,* June 23rd 2003)

ALFRED KRUPP (1812-1887)

THE CANNON KING

The astounding rise of the Krupp family is very much part of the rapid industrial growth of Germany. The recorded history of the Krupps starts in 1587 with the entry of one Arudt Krupp who ran a flourishing business in wine, grocery and real estate. Friedreich Krupp founded in 1811 the firm of Fried Krupp. He had produced, after considerable experimentation, a formula for making high quality cast steel. He left this secret and the near bankrupt factory as legacy to his 14-year-old son Alfred, when he died in 1826 at the early age of 39. This young boy was to become the real founder of the Krupp industrial empire.

Meagre education

Alfred Krupp was born at Essen, Germany on April 26, 1812. He received a meager education: at the age of 14, he had to discontinue school (1826) to run the small family steel works. With the help of his valiant mother, he quickly tightened his hold on the firm and explored ways of improving and expanding its production.

His inventions

In 1830, Krupp took up manufacture of steel rolls. He invented the spoon roll for making spoons and forks. He also specialised in making fine steel required for building new machines. In the 1840s he began introducing newly developed machinery, such as the famous giant hammer 'Fritz', by which he established the competitiveness of his steel with the English steel.

Breakthrough in steel products

His major breakthrough came in 1851, when Krupp displayed a flawless 4-ton steel ingot and his firm's first steel cannon in the 1851 World Exhibition at London. His ***production of the seamless tyre*** a year later raised his firm to the rank of one of the world's major suppliers of railroad equipment. Till then, railway axels and springs were made of only cast steel, but Krupp's seamless tyre greatly contributed to the speed and safety of railway travel. This product is still preserved in ***the firm's logo of 'three interlocking rings'.*** With the mass production of rails, the company grew into a vast enterprise with collieries, ore-mines, steel plants and blast furnaces. Krupp was ***the first to introduce the open hearth steel making process in Europe.***

Manufacture of armament

Krupp established in 1863 a physical laboratory for testing of steels. To prove the quality of his steel, he turned to making guns. Drilled out of a single block of cast steel, his gun was the marvel of its day — but acceptance of the steel cannon was slow. But it was only after the astounding performance of his guns in the Franco–German War (1870-71), did the boost in the sales of the gun begin. Though he gained his reputation through the manufacture of armament, the goods that flowed from his works were predominantly for peaceful purposes.

Krupp pioneered the vertical mixed company in Germany, by adding mining, machine building, power and transportation concerns to his industrial empire. He can be considered as the founder of modern warfare. As a recognition of his achievements in the manufacture of war weapons, he was awarded 44 military medals, stars, crosses and multiple honours by the governments of Austria, Belgium, Spain, Sweden, Russia, Turkey and Japan.

An eccentric genius and an insomniac, Krupp lay awake nights scribbling letters and industrial plans. Although he worked at forges, he was terrified of fire and built his castle of glass and stone, without a stick of wood. Because he believed that the odour of horse manure was inspiring, he built his study over his stable, with ducts to waft the scent upward. Yet, his workmen were frantically loyal to him.

Unique welfare schemes

Recognising early the human problems arising out of industrialisation, Krupp created comprehensive welfare schemes, low cost housing, consumer cooperatives, a sickness fund (1836), a pension fund (1855), churches, hospitals and housing settlements for retired workers. These institutions served as models for the social legislation enacted under Bismark (1883-1889). ***Thus he fostered the employee's pride in belonging to the company.***

Known variously as 'Alfred the Great' and the 'Cannon King', Alfred Krupp perfected the art of steel making and invented new products. At one time he had only seven employees in his steel plant. When he died in Essen on July 14, 1887, his industrial empire employed 21,000 persons.

(*The Biographical Dictionary of Scientists,* Vol. 6 Engineers & Scientists, Blond Educational, 1984)

Hugh Callendar (1863-1930)

Pioneer in steam engineering

Callendar's passion was for the study of thermodynamic properties of gases and vapours. His innovations centred around steam turbines.

History of temperature measurement

The concept of temperature as 'hot or cold' was known for a long time, but defining them quantitatively had to wait for the genius of Galileo (1564-1642) to arrive on the scene.

By observing the expansion with heat of air, water and alcohol, he got a quantitative grasp of thermal phenomena. He made an 'air thermoscope' that consisted of a glass tube of small diameter. When cooled, the air contracted moving the liquid up, when heated the air expanded, moving the liquid down. From then on, all thermometers depended on the principle that gases and liquids expand on heating. There are other properties of materials that change with temperature, which can also be employed.

Electrical resistance is one such property, that can be used for accurate measurement of temperature. It was Humphrey Davy (1778-1829) who discovered in 1821 the dependence of the electrical resistance of metals on temperature. Werner Siemens (1816-92) successfully used this phenomenon to construct a platinum resistance thermometer (1861). Hugh Callendar refined the accuracy of the above instrument, by showing that the resistance of a properly made platinum wire can be related to the reading of the normal air thermometer, by a parabolic formula that was accurate within 1%.

Education and career

Hugh Callendar was born on April 18, 1863 in Gloucestershire, England. He received his school education at Marlborough. He then attended the Trinity College, University of Cambridge, where he graduated both in classics (1884) and in mathematics (1885). He was a fellow at Cambridge from 1886 to 1888.

From 1888, he held important academic positions successively: professor of physics at Royal Holloway College (1888-93), McGill University, Montreal (1893-1902), Royal College of Science (1902-07) and Imperial College of Science and Technology. The last post he held from 1902 until his death on January 21, 1930.

Developing the platinum resistance thermometer

Callendar's main contributions were mainly in the field of experimental science, in particular, calorimetry and thermodynamics.

He designed special apparatus which ushered in new standards of accuracy in measuring the thermal properties of water and steam. Callendar's first publication in the *Proceedings of the Royal Society (1886)* dealt with platinum resistance thermometry. In a later paper (1891), he gave practical suggestions for the construction of the platinum resistance thermometer, employing mica insulation to avoid strains in the platinum wire. He later added to it an automatic recording bridge.

The platinum resistance thermometer gave the research worker a sensitive laboratory instrument for an easy and accurate method of measuring temperature and replaced the more complicated air thermometer. ***It also provided the engineer a convenient method of heat regulation, which can be readily implemented in industrial operations.***

Callendar's work got international recognition, as the Committee on Electrical Standards accepted his proposals (1899) for a standard scale of temperature based on the platinum thermometer. This scale is between –182.3°C (the boiling point of liquid oxygen) and 630.5°C (the melting point of antimony).

Callendar demonstrated (1898) its fundamental application in a number of joint studies with his assistants: the determination of the variations with temperature and strength of solution of the electromotive force of the Clark standard cell; the existence of steam in a supersaturated state, which was considered at that time of negligible importance in steam engineering.

Study of the thermodynamic problems

Callendar developed the continuous flow electric calorimeter, which almost eliminated the large corrections for heat capacity of the apparatus that is required in normal calorimetry.

The first application of this instrument was the measurement of the specific heat of water between 2ºC and 1002ºC, together with a determination of the mechanical equivalent of heat (1902). This opened a new technique for measuring specific heats of other liquids.

In 1900, Callendar first put forward his equation for an imperfect gas, as a result of his study of the thermodynamic properties of gases and vapours. This work led him to the formulation of his steam equation and the publication of the Callendar steam tables (1915), which are still widely used by engineers.

Callendar's passion was for the study of thermodynamic properties of gases and vapours, which naturally involved him in the study of engineering problems connected with steam turbines and internal combustion engines. He has won a permanent name in engineering — ***'Callendria' in steam engineering.***

Honours

Callendar was elected FRS in 1894. He received the Rumford Medal of the Royal Society (1906) and the Hawksley Medal of the Institution of Mechanical Engineers for ***his contribution to the flow of steam through nozzles*** (1915).

Callendar was a man who showed interest in non-technical fields also. He was the ***inventor of a system of shorthand*** which was used in some parts of British colonies. He published a book *A Manual of Shorthand and A System of Phonetic Spelling (1890).*

Fritz Haber (1868-1934)

Who invented the ammonia process

Early in the twentieth century, the demand for nitrogen fertiliser greatly exceeded the supply, which came mainly from Chilean nitrate. Utilising atmospheric nitrogen for this purpose engaged the attention of scientists all over the world. The first to venture the high-pressure method for synthesising ammonia was Henry Le Chatelier (1850–1936) but an explosion because of the presence of air in the gas mixture induced him to abandon his experiments. The problem posed both theoretical and practical challenges. Fritz Haber looked into several possible solutions and established the condition for large-scale synthesis of ammonia. He was awarded in 1918 ***the Noble Prize in Chemistry.***

Higher education

Fritz Haber was born on December 9, 1868 at Breslau, Germany. His father was a merchant dealing in pigments and dye stuffs. After completing his education in the local gymnasium, he attended the universities of Berlin, Heidelberg and Zurich. He obtained the Ph.D degree in 1891 from the Chariottenburg 'Technische Hochschule'.

Research work in a wide range

After spending an year in three different factories and six months in his father's business, he joined the University of Jena to take up research in organic chemistry. Chance brought him in 1893 to the Technische Hochschule at Karlsruhe, where ***he threw himself with great zeal into the teaching of physical chemistry.*** His researches in electrochemistry and thermodynamics and his book *The Theoretical Basis of Technical Electrochemistry* won him the post of professor of physical chemistry (1898). His other pioneering work *The Thermodynamics of Technical Gas Reactions* (1905) had considerable influence on teaching and research.

Although he had no formal education in the area of physical chemistry, he investigated a wide range of subjects — electrochemical reduction of nitrobenzene, fuel cells, electrolysis of crystalline salts and so on.

Gas reaction

Haber's outstanding achievement was the gas reaction, namely, synthesising ammonia from hydrogen and nitrogen. Walther Nernst (1864-1941), a leader in physical chemistry, obtained data which were at variance with Haber's. ***Haber considered this discrepancy in values a personal challenge.*** Working with one of his students, he eventually showed that a constant used by Nernst in his calculations was the cause of the erroneous values.

In 1911, he was appointed director of the Kaiser Wilhelm Institute of Physical Chemistry, Dahlem. With the outbreak of World War I in 1914, he placed himself and his laboratory at the service of his country. In 1916, he was made chief of the Chemical Warfare Service and every detail of the chemical supply and research came under his preview. By that time his process of nitrogen fixation was used in supplying Germany with nitrogen compounds, which were needed for fertilisers and explosives. He played a leading part in the development of poison gas as a weapon.

When he was honoured with the Nobel Prize, it was denounced by some French, British and American scientists. To be condemned by fellow scientists also involved in war work, deeply affected his spiritual and physical condition.

Patriotism in a practical way

During the early post-war years, Haber continued his patriotic efforts. When Germany was required to pay enormous repatriations, Haber sought to find a way of extracting gold from seawater. Several ocean trips, though they ended in a failure, were undertaken till 1926. The failure was bitter disappointment to him; still, a very accurate method of gold analysis was found.

His institute at Dahlem became the world's leading center of research, with a large and distinguished international staff. The record of his research work reflected the main developments in more than 700 publications in scientific journals of physical chemistry over 40 years. Haber accepted an invitation from the Cambridge University. After working for four months he left for a vacation in Italy, but on the way he died on Jan 29, 1934, in Basel, Switzerland.

THEODORE VON KARMAN (1881-1963)

WHO MADE FLYING FASTER THAN SOUND POSSIBLE

Karman's teaching style was uncommon. He gave graphic examples, as in the case of two monkeys hanging on either side of a rope over a pulley to explain the concept of inertia.

Guidance from his father, an educationist

Theodore von Karman was born on May 11, 1881 at Budapest, Hungary. ***He is the product of the famous Hungarian educational system***, which produced theoretical physicists like Edwar Teller and John von Neumann and control theorists Rudolf Kalman and Leo Szilard. His father was a noted educationist and founded the Minta Model Gymnasium.

Karman showed at an early age a natural mathematical ability but ***his father guided him towards engineering and also inducted in him a passion for teaching as a career.*** In later years, he was delighted when engineers to whom he had imparted his methodological approach acknowledged him as their teacher.

Influence of Klein and Hilbert

After completing his engineering education in 1902 at the Royal University in Budapest, he did military service for a year and then returned to the university to pursue his engineering career in the academic world. Thc research began here on the strength of materials prepared by him for making important contributions to the design of aircraft structures. He was awarded a two-year fellowship to study under Ludwig Prandtl (1875-1953) at the University of Gottingen. He was also greatly influenced by Gottingen's eminent mathematicians Felix Klein (1849-1925) and David Hilbert (1862-1943). Karman earned his doctorate degree in 1908, after which he accepted a position as a research assistant to Prandtl, the famous aerodynamist. From the latter, he learnt the method of abstracting the basic physical elements of a complex process and analysing it with mathematical techniques.

Karman Vortex Street

In 1911, he discovered that the oscillations of cylinder in a flow tank were caused by the alternate shedding of vortices from the top and then the bottom of the cylinder. This phenomenon, now famous as 'Karman Vortex Street' explained the collapse of the Tacoma Navrons Bridge in 1940 in the U.S.

In 1913, Karman accepted the chair of aeronautics at the Technical University, Aachen, Germany, and served till 1930, devoting his time to teaching and research. Students arrived from various countries, attracted by the intellectual environment he had created.

Career in Cal Tech

During his first visit to the U.S. in 1926, he designed a wind tunnel, the centerpiece of the new laboratory in the California Institute of Technology, Pasadena. In 1930, he was offered the post of director of the Guggenheim Aeronautical Laboratory at Cal Tech. For a time he resisted due to his love for Aachen, but the rise of Nazism drove him to accept the post. He became an American citizen in 1936.

Shortly after his arrival at Cal Tech, his laboratory became again the center of the world of aeronautical sciences. ***His scientific contributions*** continued unabated, ranging ***from the field of fluid mechanics, aircraft structure and supersonic flight to the applications of mathematics in engineering.*** He convinced the air force about the possibility of flying faster than the speed of sound and was instrumental in the decision to build the famous Bell X-1. In spite of the academicians doubting the possibilities of rocket propulsion in 1936, Karman supported the interest of a group of these students in the subject. In 1940, the prototype of the famed 'et-assisted take off rocket' was constructed. He served as director of the Guggenheim Aeronautical Laboratory (1930-49) and the Jet Propulsion Laboratory (1942-45).

When World War II began, Karman enjoyed great reputation as advisor to the army air corps. ***His famous report for the Pentagon*** *Where we stand* ***proved a prophetic and accurate analysis,*** predicting many things — supersonic combat aircraft, missiles for long-range attack, the integration of radar and communications technology into air operations.

His superb teaching style

Karman's teaching style was uncommon, revealing the fun and vitality of the man. He gave graphic examples, as in the case of two monkeys hanging on either side of a rope over a pulley to explain the concept of inertia. If one monkey climbs slowly, he overcomes his own weight; whereas if he climbs fast, he overcomes both his own weight and the inertia. So the monkey on the opposite side rises too and gets a free ride, though he is lazy and exerts no energy on the rope. (*The Wind and Beyond* by von Karman and Lee Edson, Little & Brown, Boston, 1967.) ***During his lifetime, laboratories were named after Karman*** at Cal Tech, US Air Force Development Center and NATO Institute in Belgium.

Karman was a bachelor. His sister was his manager and hostess. He was so devoted to his sister that her death in 1951 plunged him into deep depression for several months, during which he was unable to work. He died in Aachen on May 6, 1963.

Investiture by Kennedy

Karman was the ***first recipient of the National Medal of Science,*** when it was instituted in 1963. John F. Kennedy, who presided over the investiture, observing that the celebrity was suffering from arthritis, rushed to escort him down a flight of stairs to reach the podium. Disengaging himself smilingly from Mr. Kennedy's caring clasp, ***Karman remarked with a merry quip, "Mr. President, one needs your benign support when climbing and not while descending".***

Frank Whittle (1907-1996) and Hans von Ohain (1911-1998)

Who pioneered the jet engine

The birth of the jet aircraft heralded the greatest advance in aviation since 1903 when the Wright brothers made their maiden flight. The jet engine to power aircraft was ***a case of simultaneous invention,*** Whittle in England and Hans von Ohain in Germany working independently.

Whittle's career

Frank Whittle (born on June 1, 1907) recognised that long-range flight at high speed would be best effected at high altitude. He sought an alternative power plant to overcome the limitations of the piston engine. He suggested the idea of using a gas turbine to produce a propelling jet; this was in 1928.

He had joined the Royal Air Force as an aircraft apprentice (1923-26) and become an officer cadet in the college at Cranwell at that time.

His novel idea

Jet propulsion was an obsession with him; ***its principle was demonstrated by his blowing up a balloon and releasing his hold on its neck, when it zipped around the room at great speed.*** He took in 1930, his first patent, which is briefly described: the jet engine forces itself forward by taking in air and pushing it backwards. The proportion generated depends on the mass and the velocity of the air expelled. ***Whittle's engine consisted of one moving part***

— a compressor, a combustion chamber and a turbine rotating together, in contrast with the hundreds of moving parts in a piston engine.

This novel idea was in advance of its time in its material demands and the Air Ministry rejected it. Whittle's dreams of high-speed flying were rudely deflated. The RAF however, was quick to recognise his talents and deputed him to the University of Cambridge to study engineering (1934-37).

The obstacles

During this period he floated a small company, Power Jets, with his two former colleagues. The jet engine, being on the Cabinet's secret list, could not win the confidence of private financial investors. So it was difficult to raise money and Whittle lost control of his company to more powerful institutions. He could not afford the £5 required to renew his patents. He relinquished them in favour of the government. ***These obstacles took a heavy toll on Whittle's mental health.***

On 15 May 1941, the first flight using the Whittle jet engine was successfully made. Whittle watched with glee his prototype jet carry out a perfect take-off, circle and landing. "There was a war on and I was an officer, I couldn't possibly in decency have made profit out of my duties", recalled Whittle.

Recognition

However, the government recognised in 1948 his invention: Whittle was compensated with £100,000 and a knighthood. He was later decorated with the Order of Merit. Whittle died in August 1996.

Von Ohain's work

Unaware of Whittle's work, von Ohain had also designed the jet engine. Von Ohain was born on December 14, 1911. He had got the idea in the early-1930s, while studying in the University of Gottingen. He found his patron and mentor in the imaginative Ernest Heinkel, who supported the von Ohain's work from the fortune he amassed making conventional aircraft. However, Heinkel's efforts to double the thrust of his experimental engine stumbled.

However, he could enlist political favour in the air ministry of his country and won the right to design and build the first jet fighters. The first jet fighter was not pressed into service until 1944.

The Draper Prize

Both Whittle and von Ohain chose to settle in the U.S. for different reasons. Whittle retired in 1948 from the RAF for reasons of health and took a professorship in 1976 at the U.S. Naval Academy in Annapolis. ***For von Ohain, the trip began as a result of reading the stories about German boys who would stow away in ships to the U.S.*** and make their fortunes in that land of opportunity. He continued his research on jets at Wright-Petterson Air Force Base for decades and finally became professor at the University of Dayton, Ohio.

In 1991, ***the Draper Prize*** ($375,000), ***the world's richest prize for engineering,*** went to Whittle and von Ohain. The Draper candidates, the citation reads, are "individuals who come up with an idea and push it through to practice, overcoming bureaucracies and inertia and getting funds for ideas to the demonstration of a working device".

The claim to innovation clearly belongs to both the inventors, who met after the World War II and became friends in the U.S. They had ready and similar answers when asked about the secret of innovation — perseverance, sweat and ingenuity.

"The great inventions of the world are seldom made because there is a ready need for them", remarks poignantly von Ohain. (*New Scientist* Vol. 132, 5 Oct. 1991)

Karl Terzaghi (1883-1963)

Father of soil mechanics

Terzaghi had to contend with some opposition to his ideas. One example was the inability of engineers to comprehend the existence of tensile stresses larger than one atmosphere.

From mechanical engineering to geology

Karl Terzaghi was born in Prague, Austria, on October 2, 1883. He studied mechanical engineering at the Technical University in Graz, graduating in 1904. As a student, he had already become aware that mechanical engineering was not his forte. Therefore, after completing one year of compulsory military service, he returned to his alma mater for another year of study, concentrating chiefly on geology.

Study of foundation engineering

The next six years, he spent on progressional work pertaining to geology in the Karst mountains in Croatia and foundations of large buildings in Russia. He used some of the original work done during this period for his doctoral thesis; in January 1912 he received his doctorate of technical sciences from the Technical University, Graz. He spent the next two years in the U.S. studying various geological and foundation problems in that country. During 1914-16, he served in World War I.

In 1916, he was appointed professor of foundation engineering at the

Imperial School for Engineering at Constantinople. During this period ***he made an exhaustive study of all available material on earthwork and foundation engineering,*** and started devising experimental techniques in this field. In 1918, Terzaghi had to leave his post at the Imperial school; he joined the American Robert College at Istanbul, where he published the fundamental differential equation for the consolidation of clays. This was presented at the First International Conference (1924) on Applied Mechanics in Delft, Holland, and was enthusiastically received. The following year, his first major book, ***a classic in soil mechanics,*** appeared in print, which is generally accepted to be the birth of this discipline.

Migrating to the U.S.

Terzaghi was invited to MIT in the U.S. in 1925 and stayed there for four years. It was the period of his greatest creative achievement and acceptance of soil mechanics by the civil engineering profession. He had, however, to contend with some opposition to his ideas. One example was the inability of engineers to comprehend the existence of tensile stresses larger than one atmosphere in water. Another was his pioneering of mat or raft foundations in soft soils in the place of pile foundations universally adopted earlier.

In 1930, he returned to Vienna. By this time his reputation was internationally established. In 1936, he was elected the president of the first international conference on soil mechanics and foundation engineering and continued to hold that position till 1957. On outbreak of World War II, after a brief sojourn in France, he migrated to the U.S. to join the Harvard University. There he served as professor till 1956 and as emeritus professor till his death in 1963. He continued to be highly active in teaching, consultancy and publication right up to the end. In consultation work, he was careful to select such assignments only, which promised to yield valuable information. The list of these projects includes important earth dams (including ***Mission Dam, later named after him)*** landslides, foundations of buildings and bridges, permafrost problems, tunnels and subways.

Style of lecturing

Terzaghi's accomplishments were the product of many outstanding personal characteristics, including a brilliant analytical mind and an insatiable capacity for sustained mental activity. He was endowed with a warm and resonant speaking voice, which carried into the farthest reaches of a large lecture hall. Until late in life, he would present lectures lasting well over an hour, without resorting to any notes.

Honours

Among his many honours were nine honorary doctorates and the award of the Norman Medal, the highest honour of American Society of Civil Engineers, four times. Terzaghi will be remembered as one of the makers of Modern Civil Engineering, adding a whole new discipline to it.

Rudolf Diesel (1858-1913)

The man behind the 'Black Mistress'

Unlike its forerunner, the gasoline engine, which uses an electric spark to ignite the fuel, the diesel engine relies on the heat generated during a compression stroke to initiate combustion. It therefore, has a higher compression than a gasoline engine making it heavier and more expensive. It also operates at slower speeds but on less refined fuel, which is cheaper and therefore has found extensive applications in transportation (locomotives, trucks, buses), construction equipment and other heavy-duty machines. Diesel's prime model which was a single ten-foot iron cylinder with a flywheel at its base ran at Augsburg on its own power for the first time on August 10, 1893, if only for a few fleeting seconds. In those suspense-filled seconds, the pressure shot to 80 atmospheres, the highest mechanically generated and recorded at that time; and the indicator plate exploded nearly killing the inventor.

This was the ***turning point of the innovation process, which Rudolf Diesel had begun in 1885,*** and which proved the momentous potential of a still imperfect engine.

Difficult years in school

Rudolf Diesel was born in Paris on March 18, 1858 and was the only son of Theodore and Elise Strobel Diesel. Theodore was a Bavarian immigrant leather craftsman and Elise, a German-born governess and a language tutor.

Unmindful of harassment by schoolmates and being teased as a 'pretty little German pig', Rudolf won grammar school honours and at the age of 12 got admission in the Ecole Primaire Superieure, one of the most prestigious schools in Paris.

The outbreak of the Franco–Prussian War however disrupted his school career. By the French War Ministry decree of August 1870, many Germans, the Diesel family among them, were declared undesirable aliens and were deported to neutral asylum in London. Rudolf, however, got an opening as a 'deserving refugee pupil and ward' in the Royal Country Trade School in Augsburg.

His career

Here again, he suffered derison, this time as a 'pretty French pig'. Rudolf nevertheless topped his class, set an all-time scholastic record and although a non-national, got a scholarship to study in the Technical University at Munich. He attained extraordinarily high grades and was ***a protege of the renowned Professor Carl von Linde, the pioneer in mechanical refrigeration.*** During his four years at Munich, Rudolf Diesel, with his earnings as an undergraduate tutor and by making and selling second degree geometrical objects, took piano and voice training lessons, rented a piano and attended operas, concerts, art exhibitions and lectures.

Diesel joined Sulzer Machine Works of Winterthur in Switzerland as a mechanic and parts designer. Two years later he returned to Paris to join the Linde Refrigeration Enterprise, then operating in five countries, as a thermal engineer and salesman. On the 'left bank' in Paris, he made friends among artists from various nations and on the right bank, among affluent professionals and established himself as a devotee of fine arts.

Thirteen-year ordeal

Diesel married at the age of 25, Martha Flasche, a German-raised governess, who bore him two sons and a daughter. In 1885, he established his first shop–laboratory in Paris and began his 13-year-ordeal to develop his engine, what he called his 'Black Mistress', as an implement to sustain peace. ***While testing ammonia gas as a possible fuel for the engine, Diesel had a close encounter with death*** in 1888, when the gas exploded.

He perceived the near catastrophe as a war suppressor and offered to the war ministries alternatives to lethal weapons, in the form of breakable glass vials charged with contact-explosive gas and a supply of chemical fireworks so that the 'militant imbeciles who revel in competitive uproars' could feign battles. Diesel offered his proposal, when the French Patent Office turned it down, to the German Consul General to Paris, the rarely smiling Count George Munster, who laughed aloud at the idea.

After three difficult years in Paris, an impoverished Diesel had to take temporary employment again, this time with Linde Enterprises in Berlin. In late 1892, he got a German development patent on his engine following which three sponsors, the Maschinenfabrik firm at Augsburg, Sulzer and Krupp supported Rudolf. One more explosion, yet another closer encounter with death. Two years later, the epochal minute arrived on February 17, 1894.

Pioneer of the power age

During the following decade he became fabulously rich and famous, had a successful lecture tour of the U.S. and eventually lost much of his fortune. His two-volume work on social philosophy, *Solidarismus,* which sold only 200 copies, did not ameliorate his financial hardships. He was, however, famous and Diesel engines were being extensively used in electric and water plants, automobiles and trucks, marine crafts, mines and for pumping in the pipelines. The franchise, mostly international, made Rudolf Diesel a millionaire but by 1904, he was already returning to his respectable poverty due to inept financial management.

In 1912, when touring the U.S. again, Diesel's health deteriorated. He had gout and was emotionally disturbed by the darkening prospects of World War I. He, however, did not know that the oncoming year was going to be his best and also the last. He returned to Munich and was greeted by his three grandchildren, saw a deluge of new adaptations of his engine, the reopening of his beloved Munich Opera and a salon art exhibition, which Maratha Diesel arranged in his honour.

Tragic end

He set out on a brief journey to London on September 27, 1913, where he was to be honoured by the engineer's convention. Two days later, on that fateful day, he boarded a ferry enroute to England with an old engineer-friend with whom he spent a cheerful evening at dinner. Later that night, his luggage and stateroom still intact, Rudolf Diesel the inventor–philosopher, disappeared from the ferry at sea, never to be seen again, presumably drowned in the English Channel.

Leo Esaki (b.1925) and Ivar Giaever (b. 1929)

Pioneers in electron tunnelling and superconductivity

The tunneling phenomenon belongs to the direct consequence of the laws of modern physics. In a series of brilliant experiments, Esaki and Giaever have given new fundamental ideas about electrons in semiconductors and superconductors. ***They received in 1973 the Nobel Prize for Physics.*** The pioneering work of Esaki provided the direct impetus for Giaever's discovery.

Leo Esaki

Leo Esaki was born on March 12, 1925 at Osaka, Japan. He obtained a B.S. in physics in 1947 and his Ph.D in 1959, both from the University of Tokyo. His original paper on tunnel diode was published in 1958 in the Physical Review, Vol. 109.

Esaki entered the U.S. as an IBM Fellow and has been engaged in semiconductor research at the IBM Thomas J. Watson Research Centre since 1960. He has, with his colleagues, pioneered semiconductor quantum structures such as man-made superlattices, exploring a new quantum regime.

Early years

Esaki started working in 1947 in a small company at Kobe, the area of his study involving thermionic electron emission from the cathode in a vacuum tube. The decade of the 1950s was the period of evolution from the vacuum tubes to transistors. He felt extremely fortunate to have started his research career in such a period of technological innovation, where the Japanese electronic industry played an active role.

Esaki moved to Sony Corporation in 1956, headed by Morita and Ibuka. They were very active at that time and turning dreams into reality was their ideal. Esaki's work led to his thesis on tunnel diode (1957). He gave the first presentation at an International Conference in Brussels in 1958.

Historical background

Following de Broglie's hypothesis of wave–partcle duality, electrons cannot be treated merely in classical way, because they show both wave and particle behaviour. ***The waves can penetrate a thin barrier, which would be a forbidden region if we treat the electron as a particle.*** The term tunnelling refers to this wave-like property — the particle tunnells through the forbidden region.

Consider electrons in a metal moving with high velocities towards a thin insulating barrier. A certain fraction of the electrons will penetrate the barrier, causing a weak tunnells current.

Many attempts were made to observe tunnelling in semiconductors. However, theory and experiments gave conflicting results: physicists lost interest in solid state tunneling in the 1930s. ***With the invention of the transistor effect in 1947, the interest in the tunnelling process was awakened.*** Esaki, at the time working in Sony, performed some deceptively simple experiments, which proved the existence of tunnelling effect in semiconductors.

Esaki's discovery: the tunnel diode

He observed a quantum mechanical effect in which a particle like an electron has a small probability of crossing an otherwise insurmountable energy barrier. So an increased current flowed across junctions between *p*-type and *n*-type germanium, provided he made the junctions very narrow and used very high concentractions of the *n* and *p* charge carriers. ***He also discovered an unforeseen aspect of tunnelling in semiconductor junctions.***

Still narrower junctions displayed a negative resistance under certain conditions; the current across the junction would decrease with voltage.

This new aspect led to the development of an important device, called the tunnel diode or ***the Esaki diode. It was highly acclaimed in the U.S. for its novelty.***

It could be used in high frequency oscillation and amplification: for circuit switching and increasing the operating speed of electronic computers. The technical significance of Esaki's discovery was great, considering that it was made at Sony which was neither a large corporation nor a government-funded research laboratory.

The next major advance in electron tunnelling effect came in the field of superconductivity through the work of Ivar Giaever in 1960.

Ivar Giaever

He was born on April 6, 1929 at Bergen, Norway. He took a degree in mechanical Engineering in 1952 from the Norwegian Institute of Technology.

In 1956, Giaever emigrated to the USA. After working in the General Electric in various assignments as applied mathematician, he joined in 1958 its research laboratory at Schenectady. He concurrently did at Rensselar Polytechnic Institute (RPI) a theoretical thesis on ordered-disordered junctions where tunnelling in superconductivity was investigated. He obtained a doctorate degree in 1964.

On scientific discovery

Bardeen et. al. published in 1957 their theory of superconductivity. A crucial part of this theory is that an energy gap appears in the electron spectrum, when a metal becomes superconducting.

Giaever speculated that the energy gap should be reflected in the current–voltage relation in a tunnelling experiment. He studied the phenomenon through a thin sandwich of evaporated metal films insulated by the natural oxide of the film. ***The experiments showed his conjecture was correct*** and his tunnelling method soon became the standard to study the energy gap in superconductors. These experiments have confirmed the validity of the theory of superconductivity. ***The pioneering work of Esaki provided the foundation and direct impetus for Giaever's discovery.***

A headline in an Oslo paper described him: "Master in billiards and bridge, almost flunked physics, gets Nobel Prize". In his Nobel lecture at Stockholm, Giaever ***some useful insight on scientific discovery.***

"My own beliefs are that the road to a scientific discovery is seldom direct, and that it does not necessarily require great expertise. I am convinced that often a newcomer to a field has a great advantage, because he is ignorant and does not know all the complicated reasons why a particular experiment should not be attempted." (*Nobel Foundation Publication,* 1974).

ROBERT WATSON-WATT (1892-1973)

PIONEER OF RADAR DEVELOPMENT

His genius lay in devising a comprehensive system that could be operated by non-specialists under war time conditions, which turned the tide in World War II.

Early life

Watson-Watt, the youngest son of a carpenter, was born on April 18, 1892, at Brechin, Scotland. He studied physics and electrical engineering at University College, Dundee. He then became assistant to the professor of natural philosophy at Dundee, until the outbreak of World War I took him in 1915 to London, as a member of the staff of the Met office. He was asked to assist in the location of the thunderstorms for the information of aviatiors.

Physics of the upper atmosphere

This opened up fundamental research into atmospherics (the transient radio emissions from lightning discharges) at the Radio Research Station, Slough, England. Appleton (1892-1965) had begun in 1924 crucial experiments concerned with the physics of the upper atmosphere. He shot radio waves into the atmosphere to see if they were reflected and if they returned to the ground. The position of the reflecting layer at 97 kilometers, known as the ionosphere, was thus the first object detected by radio location, and it led to the birth of RADAR (radio detection and ranging).

Technique for locating enemy aircraft

In 1935, Watson-Watt put forward his memorandum to the Tizzard Committee for locating enemy aircraft by means of the radio pulse technique. He proposed transmitting short pulses covering the airspace to be explored. If the airspace contains an aircraft, a very small proportion of the energy (called echo) is reflected back to the transmitting station and can be displayed on a cathode-ray oscilloscope (CRO). ***The time delay in the return of the echo from the aircraft is used to determine the range.***

Development of radar

In this respect, there was simultaneity of discovery in several countries: but no government other than the U.K. took up the implications so quickly. The successful development of radar was a team effort with Watson-Watt as its captain. ***Perhaps more than any other in the history of science, he was in the right place at the right time.***

Watson-Watt claimed the invention of radar, but, as with other science-based innovations, its growth had been achieved through, inter alia, the developments in other fields such as CRO, direction-finding aerials, ground-scanning devices, Appleton's research work on the ionosphere.

Applications

When World War II broke out in 1938, the radar system was virtually complete. ***It played a vital role in the Battle of Britain and turned the tide from defeat to victory.*** Winston Churchill acknowledged in the House of Commons the practical importance of this invention to the Allied War Effort.

They extended to many fields of engineering: ship and aerial navigation, remote sensing, tracking artificial satellites circling the earth.

Honours

Watson-Watt earned many honours for his research: FRS (1941), Knighthood (1942), unique award of £50,000 recommended by the Royal Commission on Awards to Inventors.

Vladimir Zworykin (1889-1982)

Inventor of the modern television camera

What John Baird did towards the promotion of mechanical television system in England, Zworykin did in the United States for an all-electronic colour television system (1925).

Early education

Vladimir Zworykin was born on July 30, 1889 in Maurom, Russia. He obtained in 1912 a degree in electrical engnineering from the St. Petersburg Institute of Technology. He then went to the College de France, Paris to begin research work on X-rays under the famous physicist Paul Langevin (1872-1946), but returned to Russia in 1914. He served in the army as a radio officer.

In 1910, he came under the tutelage of Boris Rosing, Professor in charge of laboratory projects at the institute; the latter introduced Zworykin to his experiments of transmitting pictures by wire. Together, they exhibited a television system, using a mechanical scanner in the transmitter and an electronic tube in the receiver.

Landing in the U.S.

Rosing disappeared during the Bolshevik revolution (1917); but Zworykin escaped from Russia. He travelled widely and landed in the U.S. in 1919. He became an American citizen in 1924.

Then he ***worked as a book-keeper to learn English.*** In 1920, he joined the Westinghouse Electric Corporation; shortly afterwards, he enrolled as a research student at the University of Pittsburgh. He graduated with a Ph.D. degree in physics in 1926, at the age of 37.

Iconoscope and its principle

Zworykin invented the iconoscope, ***the first electronic scanning television camera.***

The principle was to focus an image on a screen made up of many photoelectric cells, each insulated, which developed a charge that depended on the intensity of the light at that point. An electron beam directed on the screen was scanned in parallel lines over the screen, discharging the photo electric cells and producing an electrical signal.

Reconstruction of the electrical signals

Zworykin used the cathode-ray tube invented in 1897 by Karl Braun (1850-1918) to produce the image in a receiver. The Braun tube (called kinescope) had an electron beam focussed by magnetic and electric fields to form a spot on a fluorescent screen. The beam was deflected by the fields in parallel lines across the screen, and the intensity of the beam varied according to the intensity of the signal. In this way, it was possible to reconstruct the electrical signals into an image. With an early version of the sytem, Zworykin managed (1923) to transmit the simple picture of a cross.

It was not until after he moved in 1929 to the Radio Corporation of America (RCA), that real progress was made towards the commercial utilisation of his invention. RCA funded Zworykin's research. He made improvements to his iconoscope, allegedly using an imaging section which was similar to the patent of P.T. Farnsworth (the farm boy to have conceived the basic principles of electronic television at the age of 13!). ***Patent litigation followed, with RCA paying Farnsworth royalties.***

By the mid-1930s, it was clear that the future of television lay with such all-electronic systems, rather than the electro-mechanical scanner pioneered by John Baird (1888-1946).

Zworykin's contributions to electron optics

He invented during World War II, several optical devices to enhance night vision. Notable among his other inventions, is the electron microscope in collaboration with James Hillier. Zworykin died at the age of 93 (July 29, 1982).

Honours

Zworykin was awarded in 1967 ***the National Medal of Science*** by the National Academy of Sciences and was named in 1977 to ***the US National Inventors Hall of Fame.***

CLAUDE SHANNON (1916-2001)

FATHER OF INFORMATION TECHNOLOGY

Shannon invented information theory. But his Ph.D topic was in the area of genetics. His thesis 'An Algebra for Theoretical Genetics' (1940) remained largely unknown. While he was doing his research, he was also pursuing the fundamental problems of communication, starting to nibble around the edges of what would later become his monumental book ***The Mathematical Theory of Communication*** (1949).

Educated at Michigan and MIT

Claude Shannon, born on April 30, 1916, in Michigan State, obtained Bachelors degrees both in mathematics and electrical engineering (1936). He then went to the Massachusetts Institute of Technology (MIT), where he obtained an MS in electrical engineering and Ph.D. in mathematics (1940).

Recognition in 2002

Shannon joined Bell Laboratories in 1941 as a mathematician and worked there till 1956. He was grateful to his employers for tolerating his work, which seemed at that time to have no practical application for telecommunications.

Ultimately, recognition came forth in 2002 by AT&T, the manufacturing arm of Bell Laboratories, ***renaming 'AT&T Research labs' as 'Shannon***

Laboratories.' From 1958 to 1978, he was Donner Professor at MIT Shannon received innumerable honours and awards, in particular ***the U.S. National Medal of Science*** (1966).

Shannon was fortunate to work under Vannevar Bush, the inventor of the ***Differential Analyser***. This device was then at the heart of analogue computers.

Though analogue computers turned out to be little more than footnotes in the history of computers, Shannon quickly made his mark with digital electronics.

Boolean Algebra to telephone switching

Shannon was not distinguished at that time as a mathematician; he was fascinated by the original work of George Boole ***Laws of Thought*** (1854).

In 1940 he published his Masters thesis, 'A symbolic analysis, of relay and switching circuits', which is recognised today as one of the most important thesis ever written.

Prize by Nobel Foundation

He showed how Boole's logical symbols could be treated as a series of ON or OFF switches and how problems could be solved by manipulating just two symbols, 1 and 0, which could be carried out by electrical switching circuits.

This was awarded ***the prestigious prize by the Nobel Foundation for the best engineering paper by an author under 30.*** (*IEEE Transactions on Information Theory,* November 2001).

Shannon's enduring fame will surely rest on his second landmark paper ***A Mathematical Theory of Communication*** (1948) published in *Bell System Technical Journal.*

The digital revolution

To a generation of telephone engineers who recognised the wave form as both the medium and the message, he came up with a radically new concept, which separated the medium from the message.

The nature of the message did not matter; it could be words, music and picture. Ultimately it was the transmission of just 1s and 0s in binary form. This was a startling notion to engineers who were accustomed to continuous waveforms.

He single handedly created today's digital revolution. Without his intuition and sustained ability to work his way through intractable problems, e-mail and the World Wide Web would not have come into existence.

Error correction codes

Shannon thought in terms of statistical ensembles and showed that if enough extra bits were added to a message, to help correct for errors, it could tunnel through the noisiest channel.

This insight as been developed over the past decades into sophisticated error correction codes. Many of Shannon's pioneering theories were ahead of their time.

They had to wait for the arrival of solid-state devices and the integrated circuit (IC) in particular (1970), for practical implementation. He tragically died of Alzheimer's disease in his 85th year (February 24, 2001).

SECTION TWO

PHYSICS

Doppler

Roentgen

L Bragg

Christian Huygens (1629-1695)

Who founded the wave theory of light

"Huygens had not the religious feeling of a Spinoza or the sensitivity of a Pascal, nor was he a philosopher of the stature of Descartes or a mathematician of the rank of Leibnitz. In an age when the human mind was making great marches into the territory of natural philosophy, Huygen's energies were thrown into the study of applied mathematics, into optical researches or astronomy and he managed somehow to pursue the most strikingly original researches in several subjects quite simultaneously. In that interesting period in Europe between the death of Galileo in 1642 and the rise to fame of Newton, Huygens, in fact, stood unchallenged as the greatest man of science of that age." — ***A.E. Bell***

Educated at home

Christian Huygens was born the son of a noted Dutch diplomat on April, 14, 1629, at The Hague. He was educated at home by a private tutor. The subjects included singing, music and Latin composition. He loved drawing and the making of mechanical models: such practical work was considered inferior to academic work, about which his tutor had misgivings. ***Rene Descartes (1596-1650) was a frequent guest in their house.*** He was impressed with some early work of Christian and he saw that great things might be expected from this serious boy.

In 1645, Christian and his elder brother Constantine studied mathematics and law at the University of Leyden where Schooten, a protege of Descartes, was the professor. Mathematics was a subject which included what we would call now mechanics, namely levers, simple machines and hydrostatics. Descartes had then published his 'Principia', a bold attempt to reduce all the changes of nature to mechanistic processes. Radical changes were taking place in people's minds and during his stay at Leyden, Christian lived in an atmosphere of intellectual ferment.

In 1647, Christian joined the college at Breda where he was taught mathematics after Schoolten, by an able Englishman, John Pell. Besides a thorough education, the most important feature of his early years was the correspondence he started with Pere Mersenne, who did much to thrash out those fundamental notions on which 17th century mechanics was based. ***Mersenne could see clearly which problems could be pursued: so Huygens was indebted to him for many of the subjects of his early research.***

Discoveries in pure and applied mathematics

Huygens worked in different areas of science, pure and applied mathematics — mechanics, which he virtually founded, optics and astronomy. His paper on hydrostatics (1650) followed his work on impact and collision. He discovered (1656) the law of conservation of momentum and was able to prove that the quantity $\frac{1}{2} mv^2$ is conserved in a collision. He studied centrifugal force and showed in 1659 its similarity to gravitational force: in fact, his gravitational theories successfully deal with several difficult points, which Newton carefully avoided.

Development of the pendulum clock

The pendulum clock that Huygens developed in 1657 made him widely known. It is true that clocks, with hands and trains of wheels driven by weights, had long been in existence: but they went irregularly, as they were only regulated by frictional resistance. Huygens was the first to realise actually the maintenance of the oscillations of a pendulum and the counting of them in a very simple manner. He substituted the pendulum in place of the oscillating body then directly coupled to the wheel of the clockwork.

This innovation raised the measurement of time, both in science and daily life, to a new level of refinement, which was hardly imagined as possible.

By 1658, major towns in Holland had pendulum tower clocks.

Huygens worked at the theory first of the simple pendulum and then of harmonically oscillating systems, throughout his life. He derived the relationship of the period, *T*, of a simple pendulum to its length, *L*. In this work, he made use of his previously worked out theory of evolutes.

A loner

These discoveries resulted in an invitation from the newly founded French Academy of Sciences, Paris. He occupied an influential position from 1666 to 1681. With the changing political climate, he gave up his post and returned to his paternal estate near The Hague. After 14 years of uninterrupted activity, he died at the age of sixty-seven (July 8,1695). He never married. ***Essentially a solitary man, he did not attract disciples and he was also slow in publishing his theories.***

Improving the telescope

The first telescopes were made in Holland early in the seventeenth century but they were imperfect. The task of improving the telescope occupied Huygens from 1652 onwards, and he had, from time to time, the collaboration of his elder brother. By means of his own telescopes, Huygens discovered Saturn's rings and also Titan, the satellite of Saturn. Using Snell's law of the refraction of light, he arrived at an understanding of the errors of lenses (spherical and chromatic aberration) and their dependence on focal length and aperture. ***He improved the Kepler's telescope, devising an eyepiece that bears his name*** and that is still in use. It consists of two convex lenses, the inner of the two working together with the objective, so that the real image is formed between the two lenses of the ocular.

Fundamental discoveries in optics

Huygens combined all known facts concerning light and supposed it to consist of a vibration, which spreads outwards from the source, just like sound in air. This is known as the ***Huygens Principle,*** which states that every point on the front of an advancing wave is itself, a source of new waves. From this principle, he published in 1678 the wave theory of light. Thomas Young (1773-1829) gave experimental support to the wave theory.

Huygens made in 1678, the important discovery of the polarisation of light which recognised that rays of light in passing through a crystal of Iceland spar or calcite are given a certain form or arrangement. This was later shown by August Fresnel (1788-1827) to be true.

A professional scientist

Huygens opposed Newton, his contemporary, younger by 14 years, over the doctrine of universal gravitation and the corpuscular theory of light. The failure to explain colour was Huygens main objection to this theory, which was for a century to overshadow his own elegant treatment of the propagation of light. He was, however, not inclined to enter into controversy.

Huygens was also a strong supporter of the atomic doctrine, which was to provide the basis for developments in chemistry. But he is not usually counted among those who developed the atomic theory, because ***his research was in the domain of mathematical physics.*** (*Christian Huygens and the Development of Science in the Seventeenth Century,* A.E. Bell, Edward Arnold & Co., London, 1947)

Modern Archimedes

Huygens was a professional scientist in an age when the boundaries of science were scarcely drawn. From 1652, he spent a lot of time on reducing to algebraic analysis problems which mathematicians such as Archimedes and Nichomedes, Pappus and Apollonius, had been able to solve only through geometry.

Huygens was helped by what Galileo (1564-1642) and Descartes (1596-1650) had done and had even surpassed their discoveries. The impact of Huygens work, in his own time and in the eighteenth century, was much less than what his genius deserved. Huygens is hailed as the 'modern Archimedes'.

Henry Cavendish (1731-1810)

Eccentric genius

He was a scientist of rare calibre, whose contributions spanned five major domains — chemistry, the atmosphere, electrical theory, theory of heat and metrology. Cavendish was reluctant to publish results which were not polished to perfection. Consequently only a fraction of his output was printed during his lifetime. It was in 1879, when Clerk Maxwell published Cavendish's papers that his genius came to light.

Assistant to his father

Henry Cavendish was born on October 10, 1731 in England. Following schooling in a seminary near London, he studied at St. Peter's College at Cambridge from 1749 to 1753 but did not take his degree. After leaving Cambridge, he became an assistant to his father whose experimental skill drew praise from the famous American scientist–statesman Benjamin Franklin (1706-1790).

In his father's laboratory, he began profound research which he pursued assiduously for more than 50 years. He was elected a Fellow of the Royal Society in 1760.

A recluse

Cavendish was a man of independent means. He spent his money entirely for scientific work and maintained a private library in his mansion open to other scientists for their use. At the age of 40 he inherited a fortune, making him one of the wealthiest men then. This did not, however, alter his lifestyle.

To the outside world, Cavendish was a shabby eccentric man who spoke little (communicating with his housekeeper through notes). He is reputed to have uttered fewer words than any other man who ever lived up to 80. He lived in solitude and retirement. Being an inveterate misogynist, female domestics were ordered to keep out of his sight. He never married and died a recluse in London on March 10, 1810.

Fundamental discoveries

Cavendish was the first to introduce quantitative techniques into chemistry. His studies on the properties of air led him to a series of fundamental discoveries: the atmosphere consists of four-fifths nitrogen and one-fifth oxygen; hydrogen is a separate substance; water is not an element but a compound that can be synthesised from hydrogen and oxygen. He discovered nitric acid and showed that it could be produced from a mixture of nitrogen, oxygen and water vapour.

Cavendish's electrical research was equally original which, if published, would have made him famous. But he shunned popular acclaim. He kept to himself a discovery, quoted in his later papers, that the force between a pair of charges is inversely proportional the square of the distance. This was subsequently established by the French scientist C.A. Coulomb (1736-1806) and later became Coulomb's law. ***He introduced the concept of electric potential, familiar in mathematics but not previously used in electrical theory*** and demonstrated that the charge on an electrical hollow sphere is lodged entirely on its surface.

Cavendish made major contributions to thermometry and specific heat of certain substances, which again remained unpublished during his lifetime.

Determination of the earth's density

His last research, known as the Cavendish experiment, related to the determination of the earth's density ***using an intricate apparatus, the torsion balance.*** It consisted of a horizontal rod suspended from its center by a long thin fibre and carrying a lead ball at each end. Near these two movable balls were two much larger fixed balls, which by their attraction upon the small ones would produce a deflection of the rod and twist the suspension fibre. Cavendish (he was nearly 70 then) calculated the force of attraction from the period of the observed oscillation and deduced the density of earth from the force as 5.84, which result was not surpassed in accuracy for a century or more.

By weighing the earth, he rendered the law of gravitation complete. The law was no longer a proportionality statement but an exact one: this was a significant addition to the science of gravitation since Newton.

Endowment

Though Cavendish lived entirely for science, he left no portion of his immense wealth to foster its growth. ***His family made amends by endowing the Cavendish Laboratory in Cambridge,*** which nurtured many eminent scientists — Maxwell, Rayleigh, J.J. Thomson, Rutherford and Lawrence Bragg.

Charles Coulomb (1736-1806)

Who laid the basis for electromagnetics

By his fundamental research in electricity and magnetism, Coulomb articulated the extension of Newtonian mechanics to empirical areas of physics and saw that the latter were raised to sophisticated disciplines in physics. The unit of electric charge was aptly named 'coulomb', in his honour.

Education and career

Charles Coulomb was born on June 14, 1736 in Angouleme, France. During his youth, the family shifted to Paris where he attended the College Mazardian and the College de France. The financial disaster that overtook his father caused Charles to follow the latter to his hometown, Montpellier; he joined the local science society in March 1757 and presented several papers in astronomy and mathematics during the next two years.

In 1760, he entered the Ecole du Genie at Mezieres and passed out in November 1761 as a military engineer. In 1764, he was posted to Martinique, West Indies, in charge of a construction project. He spent nine years here. This experience gave him the foundation for his later work — *Memoirs in mechanics*. But the stay impaired his health so badly that he was never again the same person.

Following his return to France, Coulomb served in seven different field posts in the Corps du Genie till 1781, which year marked a decisive break in his life and career. His election to the Academy of Sciences assured him permanent residence in Paris. ***Coulomb the engineer became a physicist and public servant.***

Coulomb's health, already weakened from his stay in Martinique, declined precipitously early in 1806 and he died on August 23. ***Coulomb died in poverty; it is believed that the Revolution took away most of his properties.***

Coulomb's original contributions were in the fields of applied mechanics, physics, electricity and magnetism.

Applied mechanics

His work covered structural mechanics, rupture of beams and masonry piers, soil mechanics, theory of friction and ergonomics. Though he had no special flair for mathematics, he was one of the first to use variational calculus in engineering problems. His study of friction won a prize in 1781 and brought him immediate recognition. It remained a standard of theory and experiment for a century and a half. Truly, it can be said that ***Coulomb created the science of friction.***

Coulomb was a pioneer in the field of ergonomics and he made some of the most significant contributions (1799) before the studies of F.W. Taylor, a century later. His memoirs were the first real study of useful work and fatigue in work.

The torsion balance

From 1781, when he acquired a permanent post in Paris, Coulomb shifted his research interests from applied mechanics to physics. It is observed that his emphasis on the importance of shear in structural mechanics is continued in his studies of torsion and in his ideas of coercive force in electrostatics and magnetism. His most famous papers were the memoirs on the theory of torsion (1777-84), the subsequent invention of the torsion balance and the series of seven memoirs on electricity and magnetism (1185-91). Coulomb's ***simple, elegant solution to the problem of torsion in cylinders and his use of the torsion balance in physical applications are landmarks in the history of physics.***

Coulomb was the first to show how the torsion suspension could provide physicists with ***a method of accurately measuring small forces (of less than 9 × 10^4 dynes).*** He proved that the force of torsion was proportional to the angle of twist within the range of simple harmonic oscillations. The theory of torsion paved the way for his constructing the theory of molecular interaction within solids and fluids, which laid the foundation for his work in electricity and magnetism.

Electricity and magnetism

Coulomb employed the torsion balance for many of his quantitative studies in electricity and magnetism. Thus ***he proved directly by experiment that the electric and magnetic forces vary inversely as the square of the distance;*** but he only implied that they are also proportional to the product of the respective charges or pole strengths. He never defined a unit magnetic pole strength or unlike Cavendish, a unit electric charge.

In electricity, he investigated losses due to leakage of electric charge; he formulated the theory that there are two classes of substances; perfect conductors and dielectrics. In magnetism, he created a theory based on molecular polarisation and introduced the concept of demagnetisation (namely, combinations of magnetic poles can cancel each other). The last of his famous series of seven memoirs dealt with magnetic compasses, for which study he shared the Paris Academy's first prize in 1777. It was mostly from the subject of charge distribution presented in his fifth (1788) and sixth (1790) memoirs, that Poisson formulated his theory of electrostatics in 1811.

Coulomb, as inspector general of public instruction from 1802 to 1806, played ***a conspicuous role in establishing the celebrated French system of 'lycees'.***

Research in engineering

In his two decades (1761-81) of field duty in the Corps du Genie, he took advantage of the opportunities of each posting and pursued with devotion his research work amidst demands of military life. In his memoir of 1773, he developed a generalised sliding wedge theory of soil mechanics that still remains in use in engineering practice. Coulomb ***defined and demonstrated the conduct of research in engineering through a combination of mathematics and physics.***

Thomas Young (1773-1829)

Physician, physicist and Egyptologist

Towards the end of the eighteenth century, there was a great controversy regarding the nature of light, whether it consisted of streams of particles or waves. In England, the corpuscular theory was strongly favoured because it had been advanced by Newton (1642-1727), a century before. In 1800, it was Thomas Young who reopened the debate and revived the wave theory of light propounded by Christian Huygens (1629-1695). Young's work in this respect was received by most English scientists as illogical and unscientific; it was also considered unpatriotic, as any opposition to a theory of Newton was unthinkable. In this background, Young's discovery of interference has to be viewed, which gave proof that light consists of waves.

A child prodigy

Thomas Young was born on June 13, 1773 at England, of Quaker parentage. A child prodigy, he could read at the age of two; he had read through the Bible twice by the time he was four. He was reading and writing Latin at the age of six and he had learnt at least five languages (which included Greek, French and Italian), when he reached the age of fourteen. Though he attended two boarding schools between 1780 and 1786, he was ***largely self-taught.***

In 1792, Young chose the medical profession under the guidance of his maternal uncle Richard Brocklesby, a distinguished physician at London. He studied at the universities of London, Edinburgh and Gottingen, from where he received his M.D. in 1796. He continued in 1797 his medical studies in Cambridge University for two years, to fulfill the expectations of his uncle who introduced him to such influential men as Edmund Burke and William Herschel. Here he became known as 'Phenomenon Young'. He was awarded M.B. in 1803 and M.D. in 1808.

Medical practice

When Brocklesby died in 1797, he left Young his London house, his library and a fortune. So in 1800, Young returned to London and opened his medical practice. Because of his outspoken diagnosis and his disapproval of the practices of purging and bleeding, which were then common, he was not successful as a doctor. This incidentally gave him the opportunity to attend regularly the meetings of the Royal Society, of which he became a member at the age of 21.

Young was appointed professor of natural philosophy (1801-03) at the Royal Institution, London. Though his lectures were erudite, his performance was disappointing, as they were too detailed for a popular audience. So he concentrated on his practice as a physician at St. George's Hospital, London.

Physiology of vision

While still a medical student, he recognised the mechanism of accommodation of the eye in focussing, and discovered the way in which the lens of the eye changes shape to focus on objects at different distances, near or far. He obtained in 1801 experimental proof for it. He further showed that astigmatism, from which he himself suffered, is due to irregular curvature of the cornea.

Young was the first to recognise that ***colour sensation is due to the presence in the retina of the structures which correspond to the three colours*** — red, green and violet and there is no need for a separate mechanism in the eye for every colour. Young's contribution was formulated into a proper theory of vision by Helmholtz (1821-1894). ***This three-contour theory is used in colour television and colour photography.***

Other fundamental contributions

Young announced in 1801, the discovery of interference. He allowed light to pass through two closely set pinholes onto a screen and demonstrated that bands of light overlapped with bands of darkness. He gave another instance that bright bands of fringes in Newton's rings result from light waves interfering so that they reinforce each other. Thus, he gave convincing proof that light consists of waves.

In mechanics, Young established many important concepts in a course of lectures that he delivered at the Royal Institution in 1807. Here he introduced ***the nomenclature 'energy' for the product of the mass of a body with the square of its velocity and 'work done' for the product of the force with the distance*** through which a body has moved. He defined the ratio 'stress by strain' in the elastic region of the loading of a material, which is named as Young's modulus. He published in 1805 a theory of action and accounted for the angle of contact in surface capillary tension for liquids. He used his wave theory to explain the colour of thin films such as soap bubbles.

Work on Egyptology

Turning to a completely different field, ***he penetrated the obscurity that had veiled for ages the hieroglyphics of Egypt*** through his deciphering of the Rosetta Stone which was found at the mouth of the river Nile. From 1815 onwards, Young published papers on Egyptology.

Shortly before his death (May 10, 1829), he persisted, against medical advice, in completing his Egyptian dictionary and expressed ***his great satisfaction that he had not spent one idle day in his life.***

Thomas Young was a scientist who possessed an extraordinary range of talents and was able to initiate ***important paths of investigation that others were to take up and complete.*** He is best known for his outstanding contributions in the field of optics.

Humphrey Davy (1778-1829) wrote: "Had he limited himself to any one department, he must have been first in that department. He knew so much ***it was difficult to say what be did not know".***

[The Biographical Dictionary of Scientists, Vol. 4 (*Physicists*), Blond Educational, Frederick Muller Ltd., London, 1985.]

Joseph Fraunhofer (1787-1826)

Who identified light sources by their spectra

Fraunhofer was the first physicist to investigate the dark lines in the sun's spectra, which are named after him. By studying these lines, it became possible to identify the elements in the sun's atmosphere: this was how the element helium was discovered.

Difficult boyhood

Fraunhofer was born on March 6, 1787 in a poor family at Straubing, Bavavria, Germany. With little formal education, he started working at the age of 10, in his father's optical workshop. ***An orphan by the age of 12, he became an apprentice to a dull, master glass-cutter at Munich.*** He found the work degrading and disgusting.

He had a near encounter with death, on July 21, 1801, when the workshop building collapsed. He was the only one to survive. A benevolent man heard of the accident and presented Fraunhofer a handsome amount. With the money, he obtained books on optics and immediate release from his six-year apprenticeship.

He, however, worked as a journeyman to his Munich master from November 1804 to May 1806, when he entered the optical shop of Munich Philosophical Instrument Company. By then, he had already mastered the best German textbooks on optics.

Fraunhofer combined his theoretical study with practical knowledge of glass-making, which enhanced the company's business. In 1811, he became a director.

The privations of youth and his delicate constitution hardly equipped Fraunhofer for glass blowing, which caused in 1824 the first symptoms of tuberculosis. Two years later (June 7, 1826) he died at Munich at the height of his scientific career.

Achievements

- In 1821 and 1823, shortly after Fresnel's interference phenomena received public attention, Fraunhofer published two papers, in which he interpreted certain diffraction patterns in terms of the wave theory of light.
- Fraunhofer was the first to extensively use ***diffraction grating, a device that disperses light in much the same way as a glass prism.*** He examined the relationship between dispersion and separation of wires in the grating and showed that the dispersion is inversely related to the distance between successive slits in the grating.
- He was ***the first to study the dark lines of the sun's spectrum, now known as 'Fraunhofer lines'.*** Such a study — a few of the lines had been earlier noticed by William Woolaston (1766-1828) in 1802 — to determine which elements are present in the source, is called spectroscopy. He plotted 576 of them and denoted the principal ones by the letters A to G, a system of identification still in use.
- In his time, methods of determining the optical constants of glass were crude, which limited the size and quality of lenses that could be produced. Using the lines in the solar spectrum as bench marks, Fraunhofer determined with precision the dispersion and refractive indices of various kinds of glass.
- Further, he showed how to apply the theory to the construction of lens combinations having minimum aberration.

This enabled him to develop telescopes of unsurpassed quality. The refractor he made for the Dorpat observatory and the heliometer he constructed for the Berlin observatory gave both institutions, world-class leadership in astronomy for several decades.

Honours

During the last few years of his life, he received State honours from his native province Bavaria and Denmark and the degree of Doctor of Philosophy from the University of Erlangen. In 1823, he was appointed director of the Physics Museum of the Bavarian Academy of Sciences in Munich.

Fraunhofer represents the highest order of ***combination of a theoretician and a skilled craftsman.*** The success he achieved made his name synonymous with industrial progress. In his honour, ***Fraunhofer Institutes have been established in Germany*** to bring useful technologies out of academic institutions to the industrial world.

AUGUST FRESNEL (1788-1827)

WHO ESTABLISHED THE WAVE THEORY OF LIGHT

His work on optics received scant public recognition. He maintained that not even acclaim from distinguished colleagues could compare with the joy in discovering a theoretical truth.

Educated as civil engineer

August Fresnel was born in Broglie, France, on May 10, 1788. His parents provided him early education at home. At the age of 12, he entered the local school in Caen. Two of his masters, who taught mathematics and grammar, made a lasting impression on his young mind.

In 1804, Fresnel was admitted to the Ecole Polytechnique in Paris, where he excelled in mathematics but made little progress in physics. After two years, he shifted to the technical school at Chaussees. He completed the three-year technical course in 1809 and entered the French government service as a civil engineer. He worked on road projects in various parts of the country.

Studies on diffraction

When Napoleon returned to France in 1815, Fresnel deserted his post in protest. So he was placed under house arrest. He took advantage of ***this enforced leisure to understand polarisation of light,*** about which he read in the newspaper. He also undertook experiments on diffraction. These developed his ideas on the wave nature of light and launched him on a decade of research. Till the spring of 1818, he bore heavily the yoke of his engineering duties and did research in his spare time. Thereafter, his assignments were in Paris. This eased his pursuit of work over the next few years, when he produced some important results. After 1824, his efforts slackened, as work with the Lighthouse Commission put severe demands on his time and health.

Fundamental contributions in optics

Fresnel found his motivation in an attempt to demonstrate that light is undulatory and not corpuscular. Challenging the prevailing Newtonian theory, he systematically elaborated, by a series of brilliant investigations, the concept. He initiated his investigations with experiments and proceeded, through analysis, to the theory.

The first phenomenon that Fresnel took up to explain by the wave theory hypothesis was the apparent failure of light waves to bend around a corner (known as diffraction) which was at variance with the observed behaviour of water waves and sound waves. He showed that from the wave theory there followed a slight bending and that it had to manifest itself in a succession of dark and luminous bands at the edge of the shadow. ***The mathematical formalism he proposed could predict the exact width of each of these bands.***

Fresnel did not have a lens of short focal length to verify his theory experimentally and did the job with an ingeniously suspended drop of honey! (*McGraw-Hill Encyclopedia of World Biography,* Vol. IV.)

Circular polarisation

In search of further clues to develop his theory, Fresnel turned his attention to the influence of reflection on polarised light. His efforts, summarised in a paper of November 1817, resulted in the ***discovery of an unusual modification of light, later called 'circular polarisation'.*** The novel light appeared to be symmetric about an axis drawn in the direction of its motion. Fresnel determined that these characteristics would follow, if the light were to consist of two components with mutually perpendicular planes of polarisation and a 90-degree phase difference.

Work amidst severe illness

All these results of fundamental value in physics were achieved in the face of bodily suffering, when Fresnel was haunted by thoughts of an early grave — with tuberculosis casting a shadow over his entire career. He derived his strength from a deep religious faith, which offered trust in Divine Providence. His work on optics received scant public recognition, though he was elected to the Academy of Sciences and the Royal Society, London. Fresnel, however, maintained that not even acclaim from distinguished colleagues could compare with the joy in discovering theoretical truth.

Fresnel succeeded fully in attaining his goal, the establishment of the wave theory of light. He became familiar with the work of the English physicist, Thomas Young (1773-1829) only after he was well through his own experiments. Not long after his death on July 14, 1827, at the early age of 39, ***scientific opinion shifted in favour of waves and opened up the pathway leading to the deeper insights of Clerk Maxwell (1831-79).***

Jean Charles Peltier (1785-1845)

Who invented the reverse of Seebeck effect

Peltier effect turned out to be a little more than a footnote in the history of physics, till William Thomson recognised its importance.

Apprentice in a trade

Peltier was born on February 22, 1785 at Ham, France. His father earned a living as a shoemaker. So his formal education was confined to the local schools. But, he displayed at an early age quick intelligence and mechanical skills.

At the age of 15, he was apprenticed to a German clockmaker who ill-treated him. His father then shifted him to another clockmaker in Paris. He worked in this way for six years.

In 1806, Peltier opened his own shop and married Mille Dufant. The marriage brought him a modest inheritance (1815). This was sufficient for his needs, and he retired at the age of 30.

Study of electricity

He devoted his attention to a wide range of studies and then became interested in phrenology of Franz Gall. At the age of 36, he was inspired to study anatomy, to obtain a complete knowledge of the structure of the brain. He attended a number of vivisection demonstrations, in which electricity was used to stimulate nerves. These led Peltier to the study of electricity, which he pursued till the end of his life (October 27, 1845).

Peltier's first scientific paper was presented to the Academy of Sciences, Paris, in 1830. He showed that chemical effects could be obtained from a dry pile if the surface area of the plates is sufficiently large. This work also displayed that Peltier had some understanding of the difference between current and voltage, with which electricians were to struggle for another ten years.

In 1821, T.J. Seebeck (1770-1831) had shown that temperature differences could produce electric currents. Peltier observed in 1834 the reverse effect: the passage of electricity though a junction of two different metals (antimony and copper) could produce a rise in temperature when passing in the reverse direction. The direction of heat transfer is controlled by the direction of the current.

Study of thermocouple circuits

Leopold Nobili, professor of physics at the Florence Museum, invented the thermoelectric couple (1830), by which an electric potential is created by the junction of two metals at different temperatures.

Stimulated by the above work, he constructed a thermoelectric thermoscope to measure the temperature distribution along a series of thermocouple circuits.

Peltier effect

Peltier discovered that a cooling effect could take place at one junction and heating at the other. He then confirmed this discovery by using an air thermometer. He did not pursue the effect he had discovered.

Its recognition had to wait for the thermodynamic work (Joule–Thomson effect) of William Thomson (1824-1907), which resulted in the formulation of the Second Law of Thermodynamics, equivalent to that given by Rudolf Clausius in the preceding year.

(*The Dictionary of Scientific Biography*)

Jean Foucault (1819-1868)

An outstanding experimental physicist

His first important demonstration was of the earth's rotation. Later, he invented the gynoscope, by which he gave clearer demonstration than the pendulum of the earth's rotation.

From medicine to physics

Jean Foucault (pronounced foo-ko), son of a Parisian bookseller, was born on September 18,1819. An indifferent student, Foucault began to study medicine but turned to physics, probably because of the influence of Alfred Donne, who was developing a photo-engraving process, known as daguerreotypy, in connection with his anatomy lectures at the Ecole School of Medicine.

Publishing textbooks

Donne made Foucault his assistant in the microscopy course, and subsequently the co-author of his textbook (1844). Foucault finally succeeded his master as science reporter for a well-known newspaper, where he wrote in a lively and precise style, the latest from the world of science. However, his pungent newspaper articles were also a source of hostility.

From 1844 to 1846, ***Foucault published geometry, arithmetic and chemistry textbooks for the bachelor's degree.*** Thereafter, he published only scientific papers.

Demonstration of earth's rotation

Foucault's first important experimental demonstration was of the earth's rotation. In the cellar of his house, he mounted a pendulum, free to swing in any direction and tied it at the extremity of its swing with a thread.

When the thread was set afire, the pendulum began swinging: on 8th January 1851, at 2 a.m., Foucault observed with delight the plane of swing gradually turning in the diurnal movement of the earth. This experiment, soon scaled up, was repeated during the next two years all over the world and gave rise to a tenfold increase in the number of scientific papers devoted to the simple pendulum.

In 1852, Foucault ***invented the gyroscope*** by which he gave a clearer demonstration than the pendulum, of the earth's rotation. He further showed that ***the gyro had the property of maintaining a fixed direction, similar to that of the magnetic needle.*** These experiments led to a sound understanding of rotating frames of reference. For this work, Foucault received in 1855 the Copley Medal of the Royal Society, London.

Extending the work of Young and Fresnel

Two researches, the fruit of collaboration between Foucault and Fizeau (1819-96) were of practical importance. They extended the experiments of Young (1773-1829) and Fresnel (1788-1827) to show that interference took place between rays of light, though the paths differed by several thousand wavelengths (1846). They showed in 1847 by studying the interference of heat rays from the sun, that radiant heat propagates like light. These two experiments considerably ***strengthened the wave theory of light.***

Determining the velocity of light

A comparison of the velocity of light in air and in water would be a crucial test to decide between the wave and particle theories of light. Arago (1786-1853) carried out experiments to determine the velocity of light terrestrially (earlier determinations were astronomical). He tried Wheatstone's rotating-mirror principle.

He had to abandon them due to failing eyesight and he passed on the task in 1850 to Foucault.

Foucault succeeded the same year, when he observed retardation of the velocity of light in water, giving support to the wave theory. He then saw how the rotating-mirror could be refined to measure the absolute velocity of light in a restricted space.

Foucault obtained in 1862 a value of 2.98 × 10^{10} cm per s, the first accurate measure of this fundamental physical constant. His improved apparatus was the basis for the later determination of the velocity of light by Michelson (1852-1931).

Foucault's ability to recognise fruitful lines of research was combined with an ***experimental skill of the innovator.*** He developed a regulator for the arc lamp, which made it possible for gas to be supplanted by electricity. His further improvement to the regulator brought the arc lamp into the theatre (1849). He invented the mercury interrupter (1856) to improve the performance of induction coils.

New techniques

Foucault worked to improve the design of telescopes. He introduced the modern technique for silvering glass to make mirrors for reflecting telescopes (1851). He evolved simple but accurate methods for testing and correcting the figure of both mirrors and lenses (1858).

Foucault turned toward precision engineering and electrical science. He demonstrated in 1855, the conversion of mechanical work into heat by turning with a crank, a copper disk placed between the poles of an electromagnet and measuring the heat produced in the disk.

His sparing use of mathematics condemned him as a lucky tinkerer, though none of his contemporaries could surpass him in technical and experimental skills.

Honours

Foucault worked in a laboratory setup in his home. Following the decoration of the Officer of Legion of Honour in 1851 for his pendulum experiment and the D.Sc. degree in 1853 for his thesis comparing the velocity of light in air and water, he was appointed physicist at the Paris Observatory.

Further honours followed: foreign member of the Royal Society, London (1864); the Academies of Berlin and St. Petersburg; finally member of the French Academy of Sciences.

Foucault was incapacitated by a stroke in July 1867 and, after a seven-month illness, died at Paris on February 11, 1868. His reputation spread after his death, ***as modern telescopic astronomy developed on the basis of the optical techniques he had invented.***

(*Dictionary of Scientific Biography,* Ch. Scribner's Sons, New York.)

HERMANN HELMHOLTZ (1821-1894)

WHO LAID THE FOUNDATION FOR MANY A SCIENCE

"I expected to find in him who is one of the foremost mathematical physicists of Europe, a man slightly older than myself, and I was not a little astonished when a light blond youth of girlish appearance came towards me... With respect to analytical acumen, clarity of thought and versatility, ***he surpasses every great scientist I have met;*** indeed I myself feel a little stupid in his presence", wrote Lord Kelvin (1824-1908) recalling his first meeting with Hermann Helmholtz.

Precocious boy

Born on August 31, 1821 at Postdam in Germany, Helmholtz had his early education in the local gymnasium. His father, who served in the Postdam Gymnasium as a teacher of philosophy and literature, taught him several foreign languages — French, English, Latin, Greek and Arabic. The precocious boy thrived in an intellectual atmosphere at home.

Service in the army

Though he showed early distinct interest in physics, his father did not have the money to send young Helmholtz to the university. Instead, he was persuaded to study medicine. So he joined in 1837, the Freiderich-Wilhelm Institute at Berlin, where he received free medical education after executing a bond to serve eight years in the army. He read privately the works of Laplace (1749-

1827), Daniel Bernoulli (1700-1782) and worked his way through higher mathematics. He received the M.D. degree in 1842.

In 1843, he was attached as a surgeon to the regiment. Since his duties did not occupy all his time, he improvised a temporary laboratory in the barracks where he started his experiments in physiology.

From physiology to physics

After release from his military duty in August 1849, Helmholtz joined the University of Koenigsberg as Director of the Pathological Institute. He moved in 1855 to the University of Bonn as Professor of Anatomy and continued his research into sensory physiology.

At Bonn, he was disappointed as reports went to the Minister of Education that his lecturers on anatomy were incompetent. He angrily dismissed these reports as the grumblings of entrenched medical men, who opposed his mechanistic physiological approach.

During 1849-1858, his scientific interests gravitated from physiology to physics. Thereon till 1871, he served the University of Heidelberg. This was among the most productive periods of his career.

Theory of hearing

Helmholtz won a reputation in the field of physiological optics with his invention of the opthalmascope in 1851. In physiological acoustics, he formulated the resonance theory of hearing (1858), which is considered as his greatest achievement. His great work, the massive *Handbook of Physiological Optics* (1867), was characterised by a keen philosophical insight, moulded by exact investigations in physiology and illustrated with mathematical precision.

The general theme that runs through Helmholtz's work may be traced to his rejection of 'Nature Philosophy'. Helmholtz opposed this view by postulating that all knowledge came through senses.

Helmholtz's growing scientific stature earned him in 1871 the chair of physics at the University of Berlin, a post he held for the rest of his life. He inaugurated his new position with a series of papers critically assessing the various competing theories of electrodynamic action. ***This work first brought***

Maxwell's field theory to the attention of continental physicists and inspired the later research of his illustrious pupil, Heinrich Hertz (1857-1894).

Fundamental contributions to other branches

Helmholtz made fundamental contributions to various branches of science — from physiology, optics, acoustics, mechanics, electrodynamics, hydrodynamics and physical chemistry to thermodynamics. He will be long remembered for his enunciation of the law of conservation of energy.

Helmholtz led a devoted life of unremitting labour at his research. Though his keen intellect did not wane, his health began to fail in 1885. He was prone to fits of depression, which only long vacations could cure. He died on September 8, 1894 at Berlin.

Renowned teacher

Besides the incalculable influence he exerted on 19th century science, Helmholtz was renowned for his intense activity as a teacher. He brought ***science to the common man through his brilliant and popular lectures.***

Helmholtz witnessed the final transition of the German University from a purely pedagogical institution to a temple of learning devoted to organised research. He brought to his laboratory research an analytical ability to examine the philosophical assumptions, on which science was then based. His life can be said to epitomise, in many respects, German science. For, during Helmholtz's lifetime, German science gained virtual supremacy in Europe.

JOSEF STEFAN (1835-1893)

WHO FIRST CALCULATED THE SUN'S TEMPERATURE

Stefan could establish the heat radiation as proportional to the fourth power of the absolute temperature. The theoretical proof was first given by his student Boltzmann.

Humble surroundings

Josef Stefan was born on March 24, 1835, in a village near Klagenfurt, Austria. His parents were illiterate and had a small hotel. So, he had been occupied in carrying sacks of flour to run it but made surprising progress in private studies, which allowed him to continue in the gymnasium at Klagenfurt.

In 1853, Stefan enrolled at the University of Vienna, from where ***he received the teacher's diploma*** four years later. He started publishing scientific papers covering all branches of physics. The versatility and thoroughness with which he undertook every task, contributed to his success in the scientific education of teachers at that period.

His career

He taught for three years in a school after which he was appointed ***professor of physics at the University of Vienna, where he served with distinction for 33 years till 1893.***

He was also director of the Institute of Experimental Physics founded by Doppler (1805-53) in 1850. Stefan was from 1885 vice-president of the Imperial Academy of Sciences and received numerous Austrian and foreign honours.

Stefan was a ***brilliant experimenter; his lectures were of unusual clarity and perfection.*** He was a silent and conscientious worker and regarded assisting honest scientific endeavour as a patriotic duty. He hardly ever went to scientific congresses. He became more light-hearted after he married at the age of 56. But he died just two years later (January 7, 1893).

Study of Newton's law

Stefan's most important contribution deals with heat radiation. ***Newton had stated a priori law*** "the speed of cooling is proportional to the difference in temperature between the heated body and the surrounding atmosphere". The law held good for temperature differences up to 40ºC or 50ºC only; ***physicists nevertheless considered Newton's law to be exact.***

Stefan collected and compared all the measurements published till then. He considered, in particular, Tyndall's experiments on the radiation of a platinum wire heated to incandescence by an electric current — from the weak red glow (about 525ºC) up to the full white glow (about 1200ºC). The intensity of the radiation increased almost by a factor 12. This result was examined in the light of the law propounded by Dulong (1785-1838) and Petit (1791-1820), introducing the absolute temperature scale.

Stefan–Boltzmann's law

So, Stefan could establish ***the heat radiation*** as ***proportional to the fourth power of the absolute temperature.*** The theoretical proof of this relationship was first achieved in 1884 by his student Boltzmann (1844-1906). This is now known as Stefan-Boltzmann's law of radiation. Applying his formula, Stefan could calculate and make ***for the first time a well-founded statement concerning the temperature of the sun*** (approximately 6000ºC).

Subtle problems in physics

Most of his research work dealt with intricate and subtle problems in physics:

the discovery of secondary rings in Newton's experiments; the relationship between surface tension and evaporation, which included Stefan's number; the heat conduction in gases and fluids; analogous work on the diffusion of gases; theory of alternating currents and works on acoustics.

His investigations on heat conduction of gases deserve particular mention; he devised the 'diathermometer' to ensure reliable measurements. These afforded a new means of testing the kinetic theory of gases. ***Ludwing Boltzmann,*** who succeeded Stefan as professor in 1894, ***continued his mentor's work and interpreted various aspects of thermodynamics on the basis of the kinetic theory of gases.***

(*Dictionary of Scientific Biography,* Vol. IX, Ch. Scribner's Sons, New York.)

Ernst Mach (1838-1916)

Who characterised supersonic flight

On October 14, 1947, a human being flew faster than the speed of sound for the first time in history, just 60 years after Ernst Mach observed (1887) shock waves in supersonic projectiles. Airplanes that fly at Mach 2 are commonplace today.

Educated at home

Ernst Mach was born at Turas in Austria on February 18, 1838, in a family of intellectuals. The family was voluntarily isolated in a farm, where his father did pioneering work on silkworm cultivation.

In his boyhood, Mach was weak and could develop only slowly. So he was instructed at home by his father until Mach was 14. The mornings were devoted to studies in Greek and Latin, history, algebra and geometry. The afternoons were spent in agricultural ventures.

Mach entered in 1853 directly the sixth class of the public gymnasium at Kremsier. The religious exercises displeased him, so his religious mentor recommended a handicraft or a career in business for ***the untalented student.*** But he enjoyed the geography lessons given by his teacher of natural history, F.X. Wessday, whom he recalled with gratitude at various times in later years.

After marginal performances in the gymnasium, he entered in 1855 the University of Vienna. Here he blossomed, spurred by interest in mathematics, physics, philosophy and history. In 1860, he received a Ph.D. in physics, for his thesis on electrical discharge and induction.

Mach worked as a 'privatdozent' (unpaid lecturer) in the laboratory of his professor and started on a series of experimental investigations to provide support for Doppler's principle relating change of musical pitch and optical frequency, to explain the phenomena of resonance in mechanically vibrating systems, capillary phenomena and radii of curvature of liquid surfaces.

Mach derived most of his income delivering popular lectures on optics, acoustics and psychophysics, and special lectures in physics for medical students. The latter was published as a book (1863).

It should be noted that, early in his career, he had accepted the atomic–molecular theory and the kinetic theory of gases. But later, he rejected uncompromisingly the atomic theory and observed "We have as little right to expect from it, as from the symbols of algebra, more than we have put into them, and certainly not more enlightened than from experience itself".

His teaching career

In 1864, Mach took up the post of professor of mathematics at the University of Graz. When he received no promise of resources, he used his own funds to buy equipment and to create the necessary research facilities.

By 1866, he was given the title of professor of physics. His interests had begun to shift from physics to the new discipline of psychophysics: he continued to pursue physiological and psychophysical problems on aural accommodation, the sense of time and spatial vision.

The important contribution was his discovery of the phenomenon that relates the psychological effect of spatially distributed light stimuli to visual perception.

This phenomenon, which later came to be known as ***Mach bands,*** has no physical basis, it was the subject of five papers published between 1865 and 1868. ***This was rediscovered in the 1960s.***

In 1867, Mach became professor of experimental physics, a position he occupied for the next 28 years, during which period he published over a hundred scientific papers. He moved to Vienna in 1895, assuming a teaching position in philosophy at the university. He retired in 1901, when ***he was appointed to the upper chamber of the Austrian parliament.***

In the present technological world where engineers and scientists are driven to specialisation in extremely narrow areas, it is educative to reflect on the personality of Mach, the supreme generalist. Here is a partial list of the fields he investigated: mechanics, physical optics, thermodynamics, physics of music, sugar cycle in grapes, physiology, history of science, epistemology, and literature. The eminent American philosopher ***William James summed up crisply "Mach knew everything about everything".***

Exceptional experimental skills

Between 1873 and 1893, Mach in collaboration with his students including his son Ludwig, devised optical and photographic techniques to study sound waves and the gas dynamics of projectiles.

In his celebrated paper presented to the Academy of Sciences in Vienna (1888), ***Mach showed, for the first time in history, a photograph of a shock wave in front of a bullet moving at supersonic speeds.*** That he was able to make the shock waves visible (using an innovative technique called the shadowgram) speaks of his exceptional experimental abilities. He could carry out such precise experiments involving split-second timing, without the aid of electronics — the vacuum tube had not been born. Mach was the first to understand the basic characteristics of supersonic flow.

Mach number

He pointed out the importance of the flow velocity v relative to the speed of sound a and observed the discontinuous and marked changes in a flow field as their ratio v/a from below 1 and to above 1.

J. Ackert called ***this ratio 'Mach number' in a memorial lecture delivered in 1929 at the Federal Institute of Technology (ETH)*** Zurich. The aerodynamic lift, drag and moments depend on the product of the Reynolds number and the Mach number. It is now recognised that these are ***the major governing parameters in fluid dynamics.***

Mach gave numerous popular lectures and published essays of educational importance, monographs on several topics: conservation of energy (1882), spectral investigations of musical tones (1883), sensation of motion (1883) and analysis of sensations (1886).

History of science

In 1887, Mach and Odstreil published the first of ***a series of physics textbooks,*** which went through 20 editions and was popular in Germany and Austria for 40 years.

Though a physicist by training, Mach was engaged in physiology, psychology and ***the history and philosophy of sciences.*** He was unusually sensitive to the importance of formulating problems in areas where physics, physiology and psychology intersect. Mach's contributions to the history of science exerted considerable influence on the late nineteenth and early twentieth century philosophic thought.

The branch of epistemology — a study of knowledge itself, that Mach originated was the outgrowth of his pre-occupation with physics. In his first major historical work, 'Die Geschichte and die Wurzel des Satzes von der Erhaltung der Arbeit' (1882), he sought to elucidate such ***pivotal questions listed below:***

How did we inherit our current scientific concepts and theories? Why are they given to us in the way that we have become accustomed to accept them rather than in some other form that may be logically more plausible? At any given time in history, what was considered to constitute evidence, verification, or a conclusive proof for a scientific theory?

For example, in the case of a family of mechanical principles — center of gravity, virtual work, d'Alembert's principle, Ganoo's principle of least constraint, Hamilton's principle — Mach illustrated that they are all related in such a way that, if any one is taken to be true, then all the results can be deduced by mathematical and logical reasoning.

Mach's progressive drift from physics to the philosophy of science via the history of science is nowhere more evident than in his work 'Erkenntnis and Irrtum' (1882). He wrote, "The object of science is the connection of phenomena, but the theories are like dry leaves which fall away when they have long ceased to be the lungs of the tree of science".

A memorial

In honour of his memory, a research institute has been named after him. ***The Ernst Mach Research Institute in Munich*** deals with research in fluid mechanics, experimental gas dynamics, ballistics, and high-speed photography (*Annual Reviews of Fluid Mechanics,* Vol. 15, pp 1-28, 1983, Palo Alto, California).

Mach was ***an active thinker, lecturer and writer*** up to the time of his death on February 19, 1916, near Munich. In a memorial tribute, Einstein (1879-1955) said, "I even believe that those who consider themselves to be adversaries of Mach scarcely know how much of Mach's outlook they have, so to speak, absorbed with their mother's milk".

LUDWIG BOLTZMANN (1844-1906)

WHO GAVE THE LAWS OF STATISTICAL PHYSICS

> Boltzmann's scientific stature was recognised during his lifetime. Still, it is one of the tragic ironies in the history of science that Boltzmann took his life, just before his conclusions were finally supported by discoveries in atomic physics.

Influence of Stefan

Ludwig Boltzmann was born on February 20, 1844, in Vienna in a comfortable middle-class family. Helped along by his devoted mother, he pursued his studies without any interruption. He studied mathematics and physics in the University of Vienna, where Joseph Stefan (1835-1893) was his classmate and later his teacher.

Boltzmann continued Stefan's work on radiation and obtained his doctorate degree in 1866. Shortly afterwards, he was chosen as Stefan's assistant at the university. ***Stefan urged Boltzmann to become familiar with the work of Clerk Maxwell (1831-79) on electromagnetic theory. He also taught Boltzmann English grammar,*** when the latter disclosed his total ignorance of the language.

Academic career

After working with Stefan for two more years, he was appointed, at the young age of 25, professor of mathematical physics at the University of Graz. Boltzmann's academic positions were all in Austria, except for the period spent at different times in Germany — Heidelberg and Berlin (1869-71). Munich (1890-93) and Leipzig (1900-02).

In 1873, Boltzmann became professor of mathematics at Vienna. He returned to Graz to take up the chair of experimental physics and served until 1889. He did the greater part of his work here. It appeared as if his native place was the best for him, so he returned in 1893 to Vienna succeeding Stefan as professor of theoretical physics.

Wide range of research

Boltzmann was an outstanding teacher. ***His versatility to teach both mathematical and experimental physics was astounding,*** considering that the subjects are so diverse in nature. ***All of his books originated as lecture notes.*** Boltzmann frequently developed valuable new insights, attempting to explain subjects on the elementary level. Although Boltzmann's contributions to the kinetic theory of gases and statistical mechanics were the fruits of research spread over 40 years, they account for only about half of his publications. The rest range over the fields of physics, mathematics and philosophy.

Boltzmann's greatest achievement was the development of statistical mechanics as a tool to explain how the properties of atoms (mass, charge and structure) determine the visible properties of matter (viscosity, thermal conductivity and diffusion). He worked out the general law for the distribution of energy in a system at a specific temperature and derived the theorem of equipartition of energy, the Maxwell–Boltzmann law. He also established the equation for the change of the distribution of atoms due to collisions and laid the ***foundation of statistical mechanics.***

Thermodynamics and electromagnetism

Boltzmann was one of the first European scientists to recognise the importance of Maxwell's work and became ***the most persuasive apostle of Maxwell in Germany.*** Under the impact of Maxwell's papers, he turned to a new field, electromagnetism. While in Munich, he published (1893) the two volumes of his *Lectures on Maxwell's theory on electricity and light.*

Boltzmann established the connection between thermodynamics and electromagnetism, which was exploited later in quantum theory. He came to know the work of the Italian Physicist Bartoli on radiation pressure and applying it, ***he derived the law of radiation — Stefan–Boltzmann law.*** Earlier, it was discovered experimentally by Stefan that ***radiation energy is proportional to the fourth power of absolute temperature.***

He interpreted various aspects of thermodynamics, especially the second law, on the basis of the kinetic theory of gases. He established in 1877 ***the famous law, relating entropy (S) and thermodynamic probability (W),*** which is ***inscribed on his tombstone;*** his formula is better known in the simple form which Max Planck (1858-1947) gave in 1901:

$S = k \log W$

In 1872, Boltzmann published the first report of a comprehensive experimental study of dielectrics conducted in the laboratories of Helmholtz (1821-94) in Berlin. He tested Maxwell's prediction that the index of refraction of a substance should be the geometric mean of its dielectric constant and its magnetic permeability and confirmed it for solid insulators and gases.

The tragic end

Boltzmann's scientific stature was recognised during his lifetime. He was honoured at national and international levels. He was elected FRS in 1899 and was awarded an honorary doctorate by the Oxford University.

Despite his great success as a scientist and his incomparable humour and optimism displayed in his speeches, ***he suffered from a spell of mental depression. His work on statistical mechanics was strongly attacked by many,*** who did not believe in the atomic theory and wanted to base all of physical science on energy considerations only. On account of the mounting criticism, he feared loss of his creativity. Finally, while on a vacation near Trieste, he put an end to his life on September 5, 1906, during an attack of depression.

The real cause for this tragedy will never be known. A factor may have been his feeling that the atomic theory, for which he had fought throughout his life, was being pushed into the background.

(*Ludwig Boltzmann, Man–physicist–philosopher,* Broda, OXBOW Press. Connecticut, 1983.)

WILLIAM CROOKES (1832-1919)

WHO MODELLED HIMSELF ON FARADAY

He was an experimental scientist of consummate skill. He was fortunate in having two devoted and brilliant assistants to whom he could entrust the experimental part of his research work. His paper on 'Crookes lenses' was published, when he was over 80.

Plunged into science

Willam Crookes was born on June 1832, at London, the son of a prosperous tailor. He did not attend regularly any school but seemed to have ***educated himself by his own efforts.*** He straightaway plunged into science by the scholarly introduction given by the doyen of chemists, Hofmann (1818-1892), at the Royal College of Chemistry, London.

From 1850 to 1854, Crookes served as Hofmann's assistant in the laboratory. He came to know Michael Faraday, Charles Wheatstone and George Stokes. By the influence of these eminent scientists, he turned his attention away from traditional chemical problems to optical problems in photography and to spectroscopy.

It seemed that Crookes modelled himself on Faraday, with whom he possessed in common, experimental skill, lecturing ability and an ignorance of mathematics. The friendship with Stokes came to his rescue, as the latter solved for him many mathematical problems in physics. However, the rigorous ***training that Hofmann gave Crookes remained the foundation for his***

subsequent research career.

Crookes became in 1854 superintendent of the Radcliffe Observatory at Oxford. He taught chemistry for some time and finally settled as a freelance consultant in 1856 at London. He built his own laboratory, where he carried out his research.

Diverse scientific interests

In his career, Crookes showed an amazing diversity and catholicity of his scientific interests, many of which seemed to have been motivated by the belief that all pure scientific researches would lead to financial rewards. His commercial ventures such as the sodium amalgamation process for gold extraction, utilisation of sewage and animal refuse and electric lighting did ensure him a comfortable living and support for his family of ten children. But none of them proved to be financially successful.

With the training he received in the Royal College of Chemistry, he continued to take lively interest in matters relating to agriculture. He was optimistic that chemists could solve the problem of shortage in food production by tapping the huge supplies of atmospheric nitrogen for the manufacture of fertilisers. Later Birkeland and Fritz Haber developed industrial processes for the fixation of nitrogen.

Under Hofmann's guidance, Crookes made in 1850, a study of the seleno-cyanides, using selenium ores from Germany. Encouraged by Wheatstone (1802-1875), he took up during the next decade work on photography and extended its application to the scientific recording of polarisation, astronomical objects and spectra. The announcement by Bunsen and Kirchoff in 1860, of the discovery of cesium and rubidium by spectrum analysis, gave further impetus to Crookes to pursue spectroscopy. So, everything in his home laboratory had to undergo spectroscopic examination.

When he found in March 1861 an anomalous bright green line in the selenium wastes, he confidently announced ***the existence of a new element, thallium,*** only to find that the element had been simultaneously isolated by the French chemist Lamy. However, thallium brought him recognition and got him FRS (1863). For the determination of its atomic weight, Crookes worked patiently from 1861 to 1871 to develop techniques.

Sophisticated instruments

Crookes devised (1875) a sophisticated instrument, radiometer or light mill, which consisted of "four arms, suspended on a steel point resting in a cup, so that it is capable of revolving horizontally. To the extremity of each arm is fastened a thin disc of pith, lamp-blacked on one side, the black surfaces facing the same way. The whole is enclosed in a glass tube, which is then exhausted and hermetically sealed". Its free rotation in strong light proved a financial boon to instrument makers, but the theory of its action proved very controversial. Crookes showed how the radiometer confirmed Maxwell's prediction that the viscosity of a gas was independent of its pressure.

Work on cathode rays

By attempting to determine the actual paths of lines of molecular pressure causing movement in the radiometer on the analogy of Faraday's lines of magnetic force, Crookes started to work on cathode rays, which had been, till then, the exclusive domain of German experimentalists. An electric radiometer whose vanes acted as a cathode showed that the dark space separating the cathode from the cathode glow extended from the blackened side of the vane. Since the thickness of the dark space ***(Crookes dark space)*** increased with decrease in pressure, the kinetic theory suggested that the dark space was a measure of the length of the mean free path.

Introducing into his work Faraday's concept of a fourth state of matter, namely radiant matter, ***Crookes paved the way for the fuller investigation by J.J. Thomson (1856-1940) of the behaviour of radiant matter in the discharge tube*** showing the following: it induced phosphorescence in minerals like the diamond; it caused the glass of the discharge tube to phosphoresce; its stream could be deflected by a magnet; it travelled in straight lines and was corpuscular in nature.

Other investigations

Crookes took up during the 1880s two major research topics — the technology of electric lighting with evacuated tubes and the examination of phosphorescent spectra, produced by bombarding minerals with radiant matter. Most substances gave continuous spectra but the beautiful continuous spectra produced by rare earth minerals led him to speculate in 1881 that these ores contained many unknown elements. Between 1886 and 1890, Crookes concluded erroneously that yttrium, gadolinium and samarium were complex mixtures of unknown elements.

It has happened repeatedly in the history of chemistry that, when the number of elements has been experimentally increased, there has been a simultaneous move to decrease their number by the introduction of a simplifying theory of matter. So, Crookes postulated that the idea of the chemical element had to be expanded to take into account the periodic law of elements.

In 1895, William Ramsay (1852-1916) had asked Crookes to examine the spectrum of a strange gas extracted from cleveite; in a famous telegram, Crookes had announced it was helium.

Despite his speculative powers, Crookes took a conservative view of the discovery of X-rays by Roentgen; for he could not believe that radioactive elements decayed spontaneously, which seemed to violate the law of conservation of energy. Crookes imagined that radium had the ability to select from the atmosphere those particles, which were moving more swiftly than the average, absorb some of their energy and eject them at a lower speed. This theory contravened the second law of thermodynamics and never received any support.

In 1903, Crookes developed the spinthanscope, an instrument whereby the alpha particles emitted from a sample of radioactive material were indirectly rendered visible through a simple microscope by their bombardment on a phosphorescent zinc sulphide screen. Even after the invention of more sophisticated electronic counting devices, Rutherford used this counting method for the estimation of alpha activity.

Crookes was an experimental scientist of consummate skill. He was fortunate in having two devoted and brilliant assistants; Charles Gimingham from 1870 to 1882 and James Gardine from 1882 to 1919, to whom he could entrust the experimental part of his research work. ***His paper on 'Crookes lenses' was published, when he was over 80.*** He published about 30 papers in obscure Victorian newspapers and periodicals.

Honours

He served as editor for several scientific journals, the most successful among them being *Chemical News,* which he founded in December 1859. He was knighted in 1897 and was decorated with the Order of Merit in 1910.

Science and religion

The controversial aspect of Crookes's carer was his entry into spiritualism and the battle between science and religion. He joined the Theosophical movement in 1883 and was undoubtedly the most experienced physics investigator of his time.

Crookes was active in scientific work till the end of his life (April 4, 1919). He was president of the Royal Society from 1913 to 1915, which period was marred by a degree of ill feeling from the younger generation, that he had sowed the wild oats of a genius past his allotted time.

(*The Dictionary of Scientific Biography,* Vol, III, Ch. Scribner Sons, New York.)

Johannes van der Waals (1837-1923)

From school teacher to Nobel Laureate

According to the Law of Corresponding States he discovered, the whole range of behaviour of substances can be predicted, once its 'critical point' is known. This result led to the liquefaction of hydrogen and helium.

Self-educated

Johannes van der Waals was born in November 23, 1837, in Leiden, Holland. In spite of being the son of a carpenter, he became a primary school teacher after his secondary education. Yet, his mind was fixed on much higher goals. He satisfied his quest for knowledge by studying in his spare time: during 1862-65, he attended courses in physics at the University of Leiden which did not permit him to enroll for a degree, as he was old!

Instead, he obtained the licentiate to teach mathematics and physics in high schools. Then van der Waals moved to The Hague to serve as the headmaster of a high school; at that time, a new law removing classical languages from the list of compulsory subjects for science students came into force. This enabled him to pass in 1873 the examinations for a doctorate degree. He was thus occupied with ***teaching in schools for two decades.***

His thesis

van der Waals' thesis 'On the continuity of the gaseous and liquid states' was the foundation for most of his future work. It won him great acclaim; and was translated into English, French and German. The ideal gas equation $PV = RT$ (known as Boyle's law) was derived from empirical values, assuming that the molecules of a gas were of negligible size compared to its volume. ***van der Waals modified Boyle's law, using rather simple mathematics, as***

$$\left(P + \frac{a}{V^2}\right)(V - b) = RT$$

van der Waals equation

The above equation of state, which is a cubic with regard to V, links the pressure P, volume V and absolute temperature T, where a and b are two constants determined by the volume of the molecules and their mutual attraction. He established the existence of a critical temperature below which a gas can be condensed into a two-phase system of vapour and liquid. As Clerk Maxwell (1831-1879) said, this at once put his name among the foremost in molecular science.

Then followed the most momentous part of van der Waals' research, the Law of Corresponding States (1880). According to it, the whole range of behaviour of a substance can be predicted; once its 'critical point' is known. ***This result led to the liquefaction of hydrogen (1898) and of helium (1908).*** His other achievements are the combination of the above law with the Second Law of Thermodynamics (1890).

At the age of 39, van der Waals became ***the first professor of physics in the University of Amsterdam,*** which was just then founded. He retired in 1907 and was ***succeeded to this chair by his son.*** He died in Amsterdam on March 8, 1923. van der Waals was a self-taught scientist.

Honours

Spending the prime of his career as a schoolteacher, he worked his way to the highest accolade that a scientist can obtain, namely the ***Nobel Prize for Physics,*** which he received in 1910. ***The weak attractive forces*** between atoms and molecules were ***named in his honour.***

Henri Becquerel (1852-1908)

Who discovered radioactivity

While he searched with persevearance for X-rays in phosphorescence, he stumbled on the discovery of the radioactivity of uranium.

A family of physicists

Henri Becquerel was born in Paris on December 15, 1852. His father and grandfather were physicists, both being members of the Academy of Sciences; he was born literally into the inner circle of French science.

After his early schooling, Becquerel received his science education at the Ecole Polytechnique (1872-74) and engineering education at the Institute des Ponts et Chaussees (1874-77). Even before he completed his course, he had begun in 1876 his teaching career as assistant teacher at the Polytechnique. Concurrently, he was assistant to his father at the Museum of Natural History, where he assumed the post of professor of physics on his father Edmond's death in 1891 (Edmond himself had succeeded his father Cesar Becquerel to the post). From then on, his professional life was shared among the museum, the polytechnique and the Ponts et Chaussees. In 1894, he became engineer-in-chief of a Government department, (later called the Department of Bridges and Highways) and in the next year he became professor of physics at the Polytechnique.

Early research work

Becquerel's early research was exclusively in optics: rotation of plane-polarised light by magnetic fields (1875-82), study of phosphorescent crystals under infrared illumination (1883) and absorption of light in crystals (1886-88). He obtained for these researches, his doctorate from the University of Paris (1888) and was elected to the Academy of Sciences (1889). After his research degree, he became substantially inactive in research.

Thus Becquerel, at the age of 43, was established in rank as professor at the prestigious Ecole Polytechnique but ***there was no indication then of his future research activities,*** which were to culminate in the discovery for which he is now remembered. Fortunately, he was familiar with his father's research on uranium salts, which were known to be phosphorescent when exposed to light.

Birth of nuclear physics

On learning in January 1896 about Roentgen's discovery of the X-rays, Becquerel's interest revolved around the question of whether all phosphorescent materials act as sources of similar rays. He tried other luminescent crystals and found that only those containing uranium emitted penetrating rays. Of more significance, he had shown that this was a particular property of uranium, which ***heralded the birth of nuclear physics.***

Subsequently, Marie Curie (1867-1934) and Pierre Curie (1859-1906) — who pushed on to search for unknown elements with the same property, discovered polonium and radium. With these discoveries, the field of radioactivity (a term coined by the Curies) was established. ***The scientific world's interest in the subject was niggardly:*** seven papers in 1896, two in 1897 and none in 1898.

A versatile experimentalist

Becquerel accomplished two other significant results: identification of electrons in the radiation of uranium and evidence of a radioactive transformation. By March 1900, Becquerel had duplicated the experiments of J.J. Thomson (1856-1940) to investigate the effect of magnetic fields on radiation from radium. He showed that electrons constituted a part of the radiation.

By an effective chemical procedure, Becquerel was able to remove from uranium a magnetically deviable (that is, beta) radiation. Crookes succeeded in separating from uranium the photographically active radiation, which he attributed to a substance named uranium X. Later Rutherford (1871-1937) and Soddy (1877-1956) succeeded in separating thorium X, analogous to uranium X. On the basis of these experiments, it was possible in a few months to formulate the transformation process which became the cornerstone of the theory of radioactivity.

Becquerel was a versatile experimentalist, with little taste for physical theories, which resulted in much of his research effort being wasted on observations of no great significance. But he displayed a persistent power of critical reflection, which often corrected his line of investigation. While he searched with perseverance for X-rays in phosphorescence, ***he stumbled on the discovery of the radioactivity of uranium.***

Family lineage

Becquerel received in ***1903 the Nobel Prize for Physics*** jointly with Pierre and Marie Curie. It was an appropriate division, for ***Becquerel's pioneering investigations had opened the way to the famous discoveries of the Curies.*** He was elected President of the Academy of Sciences in 1908.

His death on August 25, 1908 at Le Croisic did not signal the end of the lineage of physicists in the Becquerel family. His son Jean became the fourth Becquerel to occupy the chair of physics at the Museum of Natural History.

William Bragg (1862-1942) and Lawrence Bragg (1890-1971)

Like father, like son

Our understanding of the structure of crystals, solid-state physics (resulting in the invention of the transistor), proteins, vitamins, and such complex organic molecules as DNA is largely due to X-ray crystallography, founded by the Braggs, William and Lawrence.

Career of William

William Bragg, the father, was born in Westward, England, on July 2, 1862. He studied mathematics at the Trinity College, Cambridge, during 1881-84, under the famous teacher E.J. Routh. His distinguished performance in the Mathematical Tripos examination led to his appointment as Professor of mathematics and physics at the young University of Adelaide, Australia in 1885. He had to teach himself physics before passing on his knowledge to his students.

Finding that no apparatus was available for practical work, he apprenticed himself to a firm of instrument makers and learnt instrument design. This training prepared him for his momentous work later in 1912, ***the invention of the X-ray spectrometer.*** Although ***he never contemplated research until the age of 42,*** in a few years he gained international reputation. He was adept at picking up a subject almost casually, making an important contribution and then dropping it.

William Bragg returned to England in 1909 presumably for educating his son at Cambridge. He was professor at the University of Leeds (1915-23) and resident professor at the Royal Institution, London (1923-42). He died on March 12, 1942. He received the Order of Merit (1931) and Presidentship of the Royal Society (1935-40). He was also a popular lecturer in science and started Christmas lectures for children, which on publication, became best sellers.

Career of Lawrence

Lawrence Bragg was born on March 31, 1890 and studied at the University of Adelaide where his father was then serving. His competence in mathematics was apparent at the age of 19. In 1909, he entered the Trinity College to study physics. During the summer vacation of 1912, ***his father discussed with him a new book on the work of Max Von Laue*** (1879-1960), the German physicist who had asserted that X-rays could be diffracted by passing them through crystals.

Lawrence carried out a series of ingenious experiments for the study of crystal structures with an instrument invented by his father. He showed that X-rays could detect the pattern of atoms and molecules in crystals; this work won the ***Nobel Prize for the father and son in 1915. At 25, Lawrence became the youngest ever laureate.*** In 1965, the Nobel Foundation commemorated his golden jubilee as a laureate by inviting him to deliver the first Nobel guest lecture.

Lawrence succeeded Rutherford in 1919 as Professor at the University of Manchester. During 1919-37, he built his first research school on silicates, metals and alloys. As Director of the National Physical Laboratory, London (1937-38) he found "the engagement book, the in-tray and the list of matters requiring urgent attention to be the deadly enemies of scientific work". He relinquished the post, returned to academic life and succeeded Rutherford again, this time at Cambridge, as Cavendish Professor of Experimental Physics (1938-53). Soon, he established a second flourishing research school to study crystals and solids.

Promoting science education

Lawrence was resident professor at the Royal Institution for 13 years from 1953. As his father had done before 1940, he brought science to the non-scientist in terms that he could understand. He was ***the most successful***

salesman for science in the U.K. introducing several new activities — "weekly school lectures for children, courses for science teachers, and lectures for civil servants whose early training had not included science as a subject of study".

Lawrence was deeply concerned throughout his life with science education. ***His advice to students was wholesome:*** "Science students should have time to enjoy a full education and come to understand something of the meaning and purpose of life".

It was estimated that more than ***a lakh of children first learned of the charm and excitement of science from Lawrence's lectures.*** At an age when many scientists lose interest in research, he successfully tackled the structure of complex organic crystals, the proteins. He retired from active scientific work in 1965 at 75. He died on July 1, 1971.

Max Born (1882-1970)

Who bridged physics and chemistry

> As the founder of lattice dynamics, that is, the theory of how atoms in solids stick together and vibrate, Max Born is one of the pre-eminent physicists of this century. His celebrated work on cohesion in ionic crystals formed the bridge between the physicist's and the chemist's ways of studying crystals. For the physicists, lattice energies of crystals were of central interest and for the chemists, heats of reaction. Born showed that the ionisation potentials of the atoms could be used to compare the chemical and physical concepts. This was a landmark. — Sir Neville Mott

Early life and education

Max Born was born on December 11, 1882 at Breslau in a middle class family. As a child, he suffered very much from bad colds and asthma (which afflicted him all through his life). So his father engaged for a year a private tutor to give him lessons at home. Then, after spending two years in a preparatory school, he joined the Wilhelms Gymnasium and passed in 1900 the school leaving examination (Abitur).

The curriculum at school was centered on Latin and Greek, with mathematics as a main subject. Though he was not bad at mathematics in the lower classes, he used to feel bored as the syllabus was dominated by Euclid. But he regained his interest later after he came into contact with a good teacher, Dr. Maschke.

The latter's way of teaching inspired Born: he did not just state theorems as other teachers did but explained the whole theory to which a theorem belonged, about its history and its applications.

Born studied at the Universities of Breslau, Heidelberg and Zurich, before he settled in Gottingen in 1904 to complete his doctoral dissertation. Following his father's advice, he did not specialise but attended courses in humanities, besides mathematics and astronomy.

Impact of his teachers

His teachers were Rosanes and London. Rosanes, a pupil of Kronecker (1823-91) gave brilliant courses on analytical geometry of the plane and in space. He introduced Born to the powerful method employing the ideas of group theory and matrix calculus, which Born later could successfully apply to physical problems (first to the theory of crystal lattices, then to quantum mechanics). London gave clear and lucid lectures on definite integrals and on analytical mechanics. These mathematical lectures had such a great impact on Born that ***they completely reversed the indifference, almost distaste, to mathematics*** he had developed at school.

Amongst his class-fellows, were Lachmann who awakened his interest in astronomy and Otto Toeplitz who created a new world for him. Born had hardly heard the name of Gauss, and certainly not those of Euler, Lagrange, Cauchy and Riemann. Otto knew from his father, a schoolmaster and mathematician, their lives and characters and also the outlines of their scientific work. Born acknowledged his debt to Otto for ***the first introduction to these pathfinders in mathematical sciences.***

Announcement of Hilbert

At the University of Gottingen, he served as an assistant to David Hilbert (1862-1943). There is a fascinating account in his autobiography, of how he became very close to this dominant figure in the mathematical world then. ***It was tradition in Gottingen to select a student and ask him to write a coherent and accurate account of the lectures*** with the help of notes made in the classroom. After seeing half a dozen manuscripts, ***Hilbert announced before a vast audience, his choice for Born. Born was happy*** as, during long discussions on the manuscripts, he got a glimpse into the working of a great mathematical mind.

Born found this custom instructive and beneficial, so later on when he had gained some experience in lecturing, he copied this tradition and selected a student who could work out his lectures. His textbooks on lattice dynamics and on optics, which became classics, were thus born.

Born received his Ph.D. in 1907 at Gottingen and later was appointed 'Privatdozent' (unpaid lecturer) in physics on the merits of his paper on the relativistic aspects of the electron. He taught at the Universities of Berlin and Frankfurt am Main. In 1921, he joined the University of Goettigen as professor of physics and served for twelve years.

At Gottingen, Born inspired a generation of brilliant theoretical physicists. Among his collaborators were Pauli, Heisenberg, Schrodinger, Dirac, Oppenheimer, Hund and Jordan; each one of them made a name in their respective fields.

With the rise of the Nazis to power, he was dismissed in August 1933 from his post as professor. He had to flee Germany and spend the next three years as lecturer in Cambridge University.

School of theoretical physics

During this period he received ***an invitation from C.V. Raman*** (1888-1970), the Director of the Indian Institute of Science, Bangalore, to visit the institution for six months and organise a school of theoretical physics. ***Born sailed for India in the autumn of 1935. The rest of the crucial story is better told in his own words.***

"As I had no other job, I was willing to accept Raman's offer, namely, a permanent position at his Institute, if he could obtain the consent of the Council. Then he insisted on my attending the next faculty meeting, which had to decide on bringing my appointment before the Council.

The English professor Aston (who had joined around the same time) went up and spoke in a most unpleasant way against Raman's motion declaring that a second rank foreigner driven out from his own country was not good enough for them. I was so shaken that, when I returned home, I simply cried." (*Max Born My Life - Recollection of Nobel Laureate,* Taylor & Francis Ltd., London

1978) And this Aston was never a professor in a British university! Most tragically, The Institute lost Born and with it, the country's first flourishing school in theoretical physics.

Career in British Universities

Born was back in Cambridge in Spring 1936 as a lecturer on a small income, he was considering immigration to the U.S. Just then, the University of Edinburgh offered him the chair of theoretical physics, in which post he served for the next 17 years. He worked hard and adapted himself to the British University system of examinations, marking answer books and tutorials. On his retirement in 1953, he returned to his homeland to settle down in a town near Gottingen.

In his retirement days, Born remained active seeking the support of the German public on the dangers of nuclear weapons. He was a founding member of ***the Pugwash movement*** (of intellectuals against nuclear war) and propagated his views through lectures, broadcasts and newspaper articles.

Honours

Born received ***the Nobel Prize in 1954*** for his work on the statistical interpretation of quantum mechanics and the highest distinction, the 'Grosse Verdienst Kreuz', equivalent to the Order of Merit, from the Federal Republic of Germany. He died on January 5, 1970.

Max Planck (1858-1947)

Who introduced quantisation of energy

The beginning of the 20th century saw the development of the quantum theory, that gave birth to modern physics. The theory was announced by Max Planck on December 14, 1900 at a meeting of the German Physical Society in Berlin. Planck's discovery was to influence physics and the related sciences to a much higher degree than he would ever have imagined.

Impact of his teacher

Max Planck was born on April 23, 1858, at Kiel, Germany. He completed his early education in the local school and in 1867 entered the Maximilian Gymnasium at Munich. He gratefully recalled his teacher Hermann Mueller who taught him mechanics and astronomy. ***Mueller's explanation*** of the principle of conservation of energy made such a deep impression on him that it ***became a foundation for Planck's future work.***

Resolve to pursue physics

When young, he also displayed his prowess at playing the piano and the organ. He had a passion for music and probably gave up his idea to choose music for his career, after this advice he got from a professional musician: ***"If you have to ask, you'd better study something else!"*** But, like his teacher, Helmholtz (1821-94), he combined physics with music.

In October 1874, Planck joined the University of Munich, where he initially chose mathematics, influenced by the lectures of Gustav Banner on calculus of variations and probability theory. But soon physics attracted him, even though one of his teachers discouraged him by saying that nothing new remained to be discovered in this branch of science.

Planck was firm in his resolve to pursue physics. He spent three semesters in the University of Berlin (1877-78) where he attended the lectures of Weierstrass (1815-97), Kirchoff (1824-87) and Helmholtz (1821-94).

Planck studied, on his own, Clausius' 'Mechanische Warmrtheoie'; apparently this private study was what had finally drawn him completely into physics. His attempt to master thermodynamics led him to his thesis on the Second Law of Thermodynamics, for which he was awarded the doctorate degree in physics *summa cum laude* by the University of Munich (1879).

His career

In 1885, he started his career at the University of Kiel. At that time, Theoretical Physics had a lower status than experimental Physics. Planck first had to struggle for its recognition as a discipline. He tried in his publications to generalise his ideas to cover the theory of dilute solutions and thermoelectricity. Based on successful research on these topics, Planck was appointed in 1889 as successor to Kirchoff at the University of Berlin and the Director of the Institute for Theoretical Physics, which post was created for him. He served for 37 years as professor, ***delivering brilliant lectures and carrying out fundamental scientific work.***

Contributions

His systematic work may be grouped under four heads — ***thermodynamics, radiation theory, relativity and philosophy of science.*** The first culminated in his 'Vorlesungen' which was well-received for more than three decades as a brilliant, systematic and skilful presentation of thermodynamics. He conducted his early studies on thermodynamic irreversibility with Maxwell's electromagnetic theory of light, in the form given by Heaviside (1850-1925) and Hertz (1857-1894). He defended Clausius' form of the Second Law of Thermodynamic against gross over-simplifications.

To explain the distribution of radiant energy emitted from a black body (a body that absorbs all the radiation falling on it) at a given temperature, he postulated that energy can exist only as quanta in multiples of an elementary quantity (*h*), since known as Planck's constant.

His radiation formula, announced on December 14, 1900, introduced quanta of energy, which was a break with the fundamental principles of physics. Many physicists considered it to be an *ad hoc* hypothesis. It was only in 1913, when Niles Bohr (1885-1962) calculated for the first time the position of the special lines of a spectrum with the aid of the quantum theory, that the theory was accepted by the world of science.

Planck's third interest arose in 1905 and 1906 from the publications of Lorentz and Einstein. In 1907 he connected the principle of relativity with his quantum of action *h*.

Reflections on science and philosophy

The pursuit of general ideas dating back to 1890 really started with the lecture he delivered at Leiden in 1908, and continued in the following years with numerous reflections on the relations of science to philosophy, religion and human nature. He occupied himself with these problems in some 20 published popular lectures.

Personal calamities

He suffered several calamities in his personal life. In 1944, his home in Berlin was destroyed in an air raid, and with it he lost all his precious manuscripts and books. In the same year, his son was executed on suspicion of conspiracy to assassinate Hitler. Yet, he maintained his interest in science and a positive attitude towards life.

Honours

Planck was awarded the ***Nobel Prize in Physics in 1918.*** Planck spent the last year of his life at Gottingen, where he ***witnessed the founding of the Max Planck Institute for the Advancement of Science.*** He died on October 4, 1947 at the age of 89.

Wilhelm Wien (1864-1928)

Whose law governs radiation from objects

Wien's work contributed to the transition from classical to quantum physics. He led us to its very gates.

Facing economic crisis in boyhood

William Wien was born in a farmer's family on January 13, 1864 at Gaffken, East Prussia. He seemed destined for the life of a landowner, but an economic crisis in the family changed his career. His parents moved to a smaller farm at Darchenstein and sent him, in 1875, to the local school and later from 1880 to 1882, to the city school at Heidelberg. He had little inclination for studies, preferring to wander through the fields.

Change in his career

Encouraged by his mother, Wien enrolled at the University of Gottingen in 1882 and simultaneously at the University of Berlin. From 1883 to 1885, he worked in the laboratory of Helmholtz (1821-94) where he studied physics for the first time. In 1886, he was awarded the doctorate for his thesis on the diffraction of light by a grating.

Until 1890, he spent his time on the farm, helping his parents. August Kundt — who succeeded Helmholtz in 1888 at the University of Berlin, reinforced Wien's desire to become a farmer and advised that, as the only son, he should take over his parent's property. Fate soon decided the issue for Wien, who did not feel "capable either of buying a horse or communicating with farm workers".

Return to academic life

In 1890, when his father sold the land, Wien returned to the laboratory of Helmholtz who had become the first president of the Physikalisch–Technische Reichsantalt (PTR), newly founded in Berlin, for the study of industrial problems. Here he worked until 1896, when he was appointed professor at Aachen.

In 1900, Wien shifted to the University of Wurzberg, as successor to Roentgen (1845-1923,), the discoverer of X-rays. He spent here the next 10 years — the most successful of his scientific career. The same year, he published his book on hydrodynamics. Wien was appointed in 1920 as professor of physics at Munich, where he remained for the rest of his life. He edited *Annalen der physik* from 1906 until his death on August 30, 1928.

In his first scientific publication, Wien demonstrated (1885) that when bright light (white or coloured) struck a single metallic edge, the diffracted light was polarised parallel to the edge. Further, comparison of the colour formation in this type of diffraction with the absorption of the diffracting material yielded the complementary colour.

The displacement law

Wien's work on black body radiation must be seen against that of Boltzmann who had deduced (1884) a law expressing the total energy of such radiation. In 1893, he published a displacement law: the wavelength at which maximum energy is radiated, is inversely proportional to the absolute temperature of the radiating black body. Others, Lummer and Pringsheim, have shown that the radiation from bodies other than black bodies also obey Wien's displacement law, with the sole difference — the Wien's constant has a different value. This is one of the most important laws in the theory of heat radiation.

Thus, it became possible to determine the temperature of bodies, within narrow limits, simply by seeking the wavelength at which the radiation is greatest. The method has been applied to characterise light sources, particularly the sun. For this work, ***Wien was awarded in 1911, the Nobel Prize for Physics.***

Wien also worked on X-ray measurements. He studied in 1898, electric discharges passing through gases, and showed that they are composed of positively charged particles. He measured their deviation by magnetic and electric fields. The method used by Wein resulted, 20 years later, in the spectrography of masses. This lead to the precise measurement of the masses of various atoms and their isotopes, necessary to calculate the energies released by nuclear reactions.

Thus, Wien's work contributed to the transition from classical to quantum physics. His immortal glory was that ***"he led us to the very gates of quantum physics"*** — Max von Laue (1879-1960).

J.J. THOMSON (1856-1940)

WHO DISCOVERED THE ELECTRON

He came to physics for want of money to enroll in the engineering course. But he entered the field at a critical juncture in the history of physics, following the great discoveries of the 19th century in electricity, magnetism and thermodynamics. ***By his discovery of the electron in 1897, he opened up a new branch of science — Modern Physics.*** From the start, it rapidly gave rise to practical applications in technology and industry.

Early education

Joseph John Thomson was born on December 18, 1856 at Manchester (Britain) where his father was a bookseller. When only 14, he joined the Owens College, which, in contrast with other colleges at the time, offered some courses in experimental physics. These kindled his interest in experimental investigations.

In 1876, Thomson won a scholarship to enter the Cambridge University. He read for mathematical tripos, which at that time covered a wide range of pure mathematics as well as its applications to many branches of physics. He became Cavendish Professor for experimental physics at Cambridge in 1884 and held the post till 1919, when he resigned in favour of Rutherford (1871-1937).

Charge of the electron

At Cambridge, Thomson worked for a number of years on the electrical conductivity of gases in cathode-ray discharge tubes. By 1899, he had estimated the charge of the electron and proved it to be a universal constant. Further, he established that the atom contains only a relatively small number of electrons, which paved the way later of Rutherford's nuclear model of the atom.

From 1895 to 1914, the Cavendish Laboratory under Thomson attracted stellar talent from all over the world: this group created a record; ***seven of the scientists winning the Nobel Prize.*** Thomson worked zealously to expand the laboratory; he financed two additions to the laboratory buildings, mainly from students' fees, seeking little aid from Government or charitable corporations. A devoted staff member made a gift for the purchase of a liquid-air machine, which the laboratory needed at that time.

An outstanding teacher

He took pedagogy seriously at all levels. He exerted himself to improve science education in schools and in universities, for which he and his close friend J.H. Poynting prepared a four volume textbook *Properties of Matter* for classroom instruction. His book *Notes on Recent Researches in Electricity and Magnetism* published in 1892 is often referred to as ***the third volume of Maxwell.***

J.J. (as he was affectionately called) was an outstanding teacher and, when in good form, an unsurpassable lecturer. ***He knew the art of holding the attention of the audience, presuming neither too much nor too little on the part of the student.*** He gave regular lectures to the junior classes in the morning and to post-graduate classes in the afternoon. He never asked students set on a research career to begin by reading about work already done, lest their own views were influenced by assumptions that might be difficult to throw off later. He advised them to re-examine basic ideas, for which he considered teaching to be helpful.

Thomson was not a good experimentalist, being clumsy with his hands, but he had a ***genius for designing apparatus and guiding research.*** His splendid contributions to physics depended almost as much on the work he inspired in others as on that which he did himself.

Honours

Thomson received the highest honours in his country: Knighthood (1908), Order of Merit (1912) and presidentship of the Royal Society (1916-20). He died in Cambridge on August 30, 1940.

Thomson was awarded the ***Nobel Prize for Physics in 1906*** for his research on the conduction of electricity through gases. He showed that electrons are particles. Thirty-one years later, ***he saw his son George Thomson receive the same award*** for showing that electrons are waves.

And ever since, the idea that all particles, not just electrons, can be described by particle or wave equations has been the foundation stone of physics.

Niels Bohr (1885-1962)

Who modelled the atom

"When it came to attacking a problem, a serious problem of physics, Bohr was marvelous to watch. I always felt that he moved with the skill of a spider in an apparently empty space, judging accurately how much weight each slender thread of argument could bear. When he had explored the field, his assurance grew and his speech became vigorous and full of vivid images." — O.R. Frisch, in the book *What little I remember* (1979).

Birth of quantum theory

The discoveries of the late nineteenth century has shown that it was impossible to understand the structure of matter using the theory of classical physics, which was valid for the description of large systems. ***A new system of concepts to deal with atomic structure was inevitable: this was quantum mechanics.***

The quantum theory came close to a revolution in our thinking about the structure of matter in particular, the nature of molecular forces, the structure of metals and the radiation from atoms. It led the way to new forms of energy, materials and breakthroughs in electronics and nuclear physics. Niels Bohr was at the centre of these developments and kept it going for half a century.

Educational career

Bohr was born in a comfortable Danish family on October 7,1885. His father was a professor of physiology and a lover of science. During his school career, Bohr showed a flair for physics and mathematics: though he was not outstanding, his father expected some original contribution from him. At school, he showed his ability to inspire affection in others, an ability that later contributed to the success of the Copenhagen school he established.

In the University of Copenhagen, he had a distinguished record and he received the gold medal. For his master's and doctorate degree, Bohr studied the electron theory of metals. This work showed him the limitations of classical theory in describing electron behaviour and the need for a radically different theory for atomic systems.

Work at Cavendish Laboratory

In 1911, Bohr visited the Cavendish Laboratory at Cambridge, he had just finished his dissertation derived from Thomson's model and hoped to discuss the problems with this doyen of atomic physics and publish his thesis in English. But J.J. Thomson (1856-1940) was not responsive, as he had ceased to work on the theory of metals. Bohr was frustrated during his first few months in England till he met Rutherford (1871-1937), who gave him all the attention he hungered for and soon invited him to Manchester.

The Rutherford–Bohr collaboration, as the prelude to the next epoch-making years in atomic physics was the background for Bohr's lifelong sense of fraternity with British scientists. It was not long before Rutherford, on being asked a theoretical question by another colleague, replied 'ask Bohr'. Soon whenever a question arose, ***the saying went around in the laboratory: 'ask Bohr'.***

It was here that he made (1912) several important contributions to atomic physics: ***he laid the basis for radioactive transformation*** and developed a theory for the energy loss of alpha particles, as they passed through matter, using the Rutherford model of the atom; and he worked on the quantum description of atomic structure.

His new theory

Bohr published, in 1913, his new theory in three papers in the *Philosophical Magazine*. The trilogy earned him early reputation. The Bohr atom, though scientifically superseded for half a century, still retains its image as a symbol of physics. The work on the quantum orbits of the hydrogen atom became classical in physics literature. ***It marked the beginning of a decade of new insights*** when quality (the specificity of chemical substances) was replaced by quantity (the number of electrons per atom).

In 1916, Bohr returned to the University of Copenhagen where he was appointed to the Chair of Theoretical Physics, a post specially created for him. Within a few months, Hendrik Kramers (1894-1952), a young Dutchman, joined the staff and became one more life-long friend of his. The two planned and ran the new University Institute for Theoretical Physics (later to become the Niels Bohr Institute). It opened in 1921 and the list of visitors over the next two decades reads like ***a roll-call of the founders of quantum mechanics:*** Pauli, Heisenberg, Dirac, Ehrenfest, Block, Casimer, Landau, Franck, Frisch. ***The ideas spurred by Bohr had enormous impact on physics and chemistry.***

In 1922, Bohr contributed several papers on the theory of atomic structure and the periodic system of elements. In the same year, be was awarded ***the Nobel Prize in Physics.***

In Germany, Bohr's atomic model was little read initially. How much courage, independence and grasp on essentials he had, is shown by the slowness with which his theory was accepted by the great body of physicists.

Work in nuclear physics

Beginning in 1934, Bohr made his major contribution through his work in nuclear physics: in particular, his theory of the compound nucleus, and his elaboration of the liquid-drop model. This work reached its climax in 1939 in a joint paper (with John Wheeler) on the theory of nuclear fission. In 1943, when the occupation of Denmark became more oppressive, he escaped to Sweden and then, via Britain, reached the U.S. He joined a large group of scientists in Los Alamos to work on atomic weapons and faced this squarely as a necessity.

In the autumn of 1945, he returned to his institute. For the next ten years, he planned the Danish Atomic Energy Commission's research establishment. He lent his name and support to the creation of Center for European Cooperation in Science (which was later to become CERN, Geneva). ***Bohr was actively engaged in a one-man campaign*** to persuade the leading statesmen of the world ***for an international understanding on nuclear weapons.***

His students and members of the Bohr institute planned to celebrate the golden jubilee of his work on atomic structure; but before this could take place, Bohr passed away on November 18, 1962, when he lay down for a brief rest in the afternoon.

Robert Millikan (1868-1953)

Who measured the charge of the electron

He decided to apply the electric field in a direction opposite to the force of gravity, balance it and suspend the electron-laden droplets in mid-air.

Learning physics by teaching

Robert Millikan was born on March 22, 1868 at Illinois, U.S.; he was the son of a congregational preacher. He received private tuition at home, besides enjoying life in the countryside, farming and fishing.

Millikan straightaway entered in 1886 the preparatory department of the Oberlin College. He had learnt next to nothing about science. So when he was asked by his Greek professor to teach the elementary physics course, it came as a complete surprise to him.

The professor's remark that "anyone who can do well in my Greek can teach physics" gave him the impetus to accept the challenge. Millikan started the preparation with an elementary textbook and pursued a Course of self-instruction in Silvanus Thomson's 'Dynamic Electric Machinery'. On being awarded M.A. for this achievement (1893), he entered the Columbia University. He was ***impressed by the lectures of Michael Pupin, who observing Millikan's academic flair, financed him for postgraduate study in Europe.*** He received his Ph.D. in 1895 and then visited the universities of Paris, Jena, Berlin and Gottingen. He heard Poincare's lectures in Paris and took a course under Max Planck (1858-1947) at Berlin.

In 1896, with the excitement of the discovery of X-rays and radioactivity fresh on his mind, Millikan started his career in the University of Chicago, as an assistant in physics.

American physics was then in its infancy. So Millikan divided his activities equally between research and the writing of text books. He held the conviction that ***the most effective way of learning physics is by problem-solving*** and not by the passive method of lecturing which he viewed as a hold-over from pre-printing days. He wrote or co-authored ***several influential text books*** — in particular, *First course in Physics* (1906) compatible with his educational philosophy. They sold steadily through the years. Because of his outstanding pedagogical achievements, Millikan was promoted (1907) to the rank of an associate professor. Eager to establish his name in research, he decided in 1908 to shelve the writing of text books and switch over to laboratory work.

Oil drop experiment

Millikan chose the problem — the measurement of the charge of electron, as no one had yet obtained a reliable value for this fundamental constant. He intuitively sensed the sources of errors in the technique developed by Townsend, Thomson and Wilson (1903) at the Cavendish Laboratory. ***Millikan quickly recognised that the evaporation at the surface of the cloud of water vapour confused the measure of its rate of fall.*** These droplets were allowed to fall, under the influence of gravity alone or gravity plus an electric field. Millikan decided to apply the electric field in a direction opposite to the force of gravity, balance it, and suspend the electron-laden droplets in midair. With the switching-on of the electric field, a few individual drops remained within the field of view. Millikan provided the key to the problem, namely observe single droplets using this balancing field technique.

Later the water droplets were replaced by oil droplets which could pick up electrons by passing through air ionised by X-rays. In 1913 he published the value for electronic charge — 4.77×10^{-10} e.s.u., which stood for two decades.

Validity of the photoelectric equation

Off and on, Millikan gave his attention to Einstein's photoelectric equation. During a visit to Berlin in 1912, he was forcefully reminded of the chaotic experimental situation regarding the famous equation. He was familiar with the experimental difficulties from investigations he had carried out in 1907, still it took him three years of intensive work (1912-1915).

Drawing on an accidental observation, Millikan discovered that alkali metals are sensitive. That was the key to the problem. Millikan employed a Faraday cage of well-oxidised copper netting, rigging up an electromagnetically operated knife to clean the metal surface. As the result of these investigations, Millikan confirmed (1915) the validity of Einstein's equation in every detail. His work also provided the best measure of Planck's constant (h) then available.

Honours

These two achievements won for Millikan the election as president of the American Physical Society (1916-1918) as well as many prizes, medals and other honours. He was awarded the ***Nobel Prize for Physics in 1923.***

Millikan participated in the war effort in Washington (1917-18): most of his activities centered on the development of submarine detection and destruction devices, to break the back of the German U-boat menace.

At Cal Tech

In 1921, he took over as director of the recently established Norman Bridge Laboratory at the California Institute of Technology (Cal Tech). Here he fostered a wide variety of research work, on everything from earthquakes to pure mathematics, but he himself took the greatest interest in the phenomenon known as 'field emission', the radiations discovered in 1912. ***He coined the term 'cosmic rays',*** a name retained to this day. At the zenith of his powers, he was America's foremost experimental physicist. He attracted and inspired a band of capable students, many of whom subsequently became his colleagues.

New educational policies

At Cal Tech, Millikan found a unique opportunity to ***implement his ideas on American education.*** Under his guidance, it grew from obscurity to a position of pre-eminence. The major educational policies he implemented were two fold: first, ***adequate emphasis on the humanities*** and second, ***close ties between basic sciences and the engineering disciplines.*** He retired at the age of 78 in 1946, but remained active as a public lecturer and spoke frequently on science and religion.

Millikan died on December 19, 1953 at Pasadena. "The secret of his success", wrote his friend, "lay to a large extent in the simple virtues instilled in his upbringing. He had a singleminded devotion to all that he was doing, and he put his work above his personal desire and aspirations."

(*The McGraw Hill Encyclopedia of World Biography,* Vol. 7)

Wolfgang Pauli (1900-1958)

Who quantised electrons

Through Sommerfeld's lectures, Pauli became acquainted with the quantum theory of the atom. He recalled in his Nobel lecture: "It was at the University of Munich that I was introduced by Sommerfeld to the structure of the atom, somewhat strange from the point of view of classical physics. I was not spared the shock which every physicist accustomed to the classical way of thinking experienced when he came to know of Bohr's postulate of quantum theory for the first time."

Choosing theoretical physics

Wolfgang Pauli was born in April 25, 1900, in Vienna, Austria. In high school, he was proficient in higher mathematics for he studied on his own Jordan's 'Cours d'analyse'. Einstein's papers on general theory of relativity, which Pauli read secretly during dull classroom hours, made a deep impression on his mind. It was, he said, as if scales had fallen from his eyes; one day, he felt in a flash he had understood the general theory of relativity.

As he decided to study theoretical physics, Pauli entered in 1918 the University of Munich to study under Arnold Sommerfeld (1868-1951), the legendary figure of theoretical physics in Germany, whose pupils included many outstanding theoreticians such as Heisenberg and Bethe. In this scientific atmosphere, Pauli perfected his analytical skills. ***Even as a student, he got a splendid opportunity to show his art of presenting science.***

When Sommerfeld was invited to contribute an article on relativity theory for the 'Mathematical Encyclopedia' (edited by Felix Klein), he intuitively chose Pauli for this important task. Pauli produced a 250-page monograph, which critically presented the mathematical foundations of the theory and its physical significance. This has remained to the present day a first-rate classical document on space, time, and matter. Sommerfeld expressed his admiration, in a letter to Einstein, that Pauli's essay was outstanding.

Discovery of the exclusion principle

In 1922, Pauli obtained his doctorate and soon thereafter began to work on the anomalous Zeeman effect. These studies culminated in the discovery of his famous exclusion principle. In its simplest form, this principle states that in a given atom, no two electrons may share the same values for all the four quantum numbers. It won him the ***Nobel Prize in 1945.*** Pauli had solved an intricate problem. By the time the article was published in the 'Handbook of Physics' (1926), new developments had taken place. The works of Heisenberg, Dirac and Schrodinger on quantum mechanics had appeared on the scene in rapid succession. So Pauli added an article on wave mechanics in the second edition (1933).

The same article was published 25 years later, unchanged, in the new edition (1958) — an unparalleled accomplishment of Pauli in the art of presenting science.

After serving at the University of Hamburg (1924), Pauli shifted to Gottingen to work as an assistant to Max Born (1882-1970). His incredible quickness and acuteness of perception enabled him to do a collosal amount of work without much effort. He worked on atomic problems and helped Max Born in directing the work of research students.

Teaching career at ETH

In 1928, the Federal Institute of Technology (ETH), Zurich, elected Pauli to the chair of theoretical physics as Debye's successor. His mentor, Sommerfeld, advised him to accept the position in order to breathe in a different scientific atmosphere. He remained in this post till his death on December 14, 1958.

Pauli did not enjoy the reputation as a good lecturer. For, he had the tendency while lecturing to reflect over the subject, which hindered teaching. And so his lectures were difficult to follow. Nevertheless his students were greatly stimulated, as ***he radiated a strong personal force. He was truly a teacher who helped in the birth of ideas, in the Socratic tradition.***

Pauli postulated in 1931, the neutrino hypothesis, namely the simultaneous production in beta decay of an unchanged particle of extremely low mass (neutrino). This was not detected experimentally until 1956. This is one of Pauli's most significant contributions to physics.

Understanding scientific creativity

Theoretical physics surely appears quite rational, but it rises from irrational depths. Pauli had experienced this, deep within himself and so took up the ideas of the Swiss psychologist Carl Jung (1875-1961) to understand the meaning of scientific creativity. These efforts resulted in the book 'Naturerklaerung and Psyche' which he published in 1952 with Jung.

Just as Pauli received a shock when he first became acquainted with the strange laws of quantum theory, so did he receive a shock from the non-conservation of the energy in weak interactions, as in beta decay. By his research work, Pauli showed how apparently paradoxical phenomena could be explained through a harmonious extension of the existing theory. And so ***theoretical physics since Kepler, Galileo and Newton appeared to him as a great house,*** the foundations of which would never be shaken.

Werner Heisenberg (1901-1976)

Who set the limits to accuracy

Max Born was the centre of a brilliant group (of theoretical physicists), which suddenly saw its way to formulate a set of laws for the behaviour of inanimate matter. One of its members who helped to establish the modern science of quantum mechanics was Werner Heisenberg.

A significant episode

Born on December 5, 1901 in Wuerzburg, Germany, Werner Heisenberg studied physics at the University of Munich. He worked on his doctoral dissertation 'Turbulence of fluid streams' under the guidance of Arnold Sommerfeld (1868-1951).

There is an interesting account given by Max Born, in his autobiography, on how Heisenberg almost failed in the oral examination. Wilhelm Wein (1864-1928), the famous physicist who was the examiner, asked Heisenberg about experimental techniques, when he got no satisfactory reply, he declared the candidate failed. After an animated dispute, Sommerfeld procured the lowest grade 'rite' for his outstanding student. The significance of this episode became apparent only a few years later.

For, one of Wien's questions was concerned with the resolving power of optical instruments due to the finite wavelength of electromagnetic radiation. The now well-known Uncertainty Principle, expressing indeterminacy of the position and momentum simultaneously, is not different from the translation into particle language of the optical formula, which says that the indeterminacy in position will be approximately equal to the wavelength. Among the questions which Heisenberg failed to answer in the oral examination, ***the problem of resolving power impressed on him as fundamental.*** When the time came, he remembered it, and the result was the principle, which made him famous not only in physics but also in philosophy.

Mathematical formalism of matrix calculus

After his doctorate in 1923, Heisenberg studied for a year at the University of Gottingen under David Hilbert (1862-1943) and Max Born (1882-1970). The next year he spent in Copenhagen working under Niels Bohr (1885-1962). No sooner had he completed his stay in Copenhagen than he solved a major problem: he devised a method of calculating the energy levels of atomic oscillators.

He was then recovering after an attack of hay fever and was inclined 'to throw his paper into the flames'. But, happily for science, that did not happen; he sent a copy to Wolfgang Pauli (1900-58) who commented on it favourably. He showed it, on returning to Gottingen in June 1925, to Born who immediately recognised its importance and sent the paper to *Phykalische Zeltshrift* for publication.

Prior to Heisenberg's work, matrices had rarely been used by physicists. ***Born identified the strange mathematical formalism in Heisenberg's paper as matrix calculus,*** the rules of which were already known to him taught by his professor Rosanes at the University of Breslau. Heisenberg knew at that time very little of matrices and had to learn them. "Some considered our suggestion as a mild form of madness", wrote Born.

The Uncertainty Principle

In 1926, Heisenberg was appointed lecturer in theoretical physics at the Copenhagen University. It was here that he gave serious thought to the discrepancy between the two formalisms of quantum theory, one based on matrix calculus and the other on wave equations. This led him to formulate the Uncertainty Principle, which won him the ***Nobel Prize in 1932. He was disturbed by the award, as his collaborator Born was left out.*** (The latter received the prize in 1954.) Heisenberg wrote a moving letter and acknowledged his debt to Born for this distinction.

His career

From 1927 to 1941, Heisenberg was professor at the University of Leipzig. For the next four years, he was director of the Kaiser Wilhelm Institute for Physics in Berlin. During World War II, he worked on the development of a nuclear reactor. After the war, he organised and became director of the Institute for Physics at Gottingen, later relocated to Munich in 1958 as Max Planck Institute.

Simply stated, the uncertainty principle ruled that no quantum mechanical system could simultaneously possess an exact position and exact momentum. That is, on the subatomic level, events cannot be predicted exactly; only their statistical properties can be determined. Hence, ***Newton's laws*** concerning the motion of bodies in space and time ***do not fit the basic processes within the atom; the traditional idea of scientific causality needs broadening*** to predict the behaviour of sub-atomic particles.

A new concept

In his papers dealing with quantum mechanics, Heisenberg gave ***a new concept of measurement process,*** emphasising the active role of the scientist. The scientist in the act of making measurements interacts with the observed object and thus causes it to be revealed not as it is in itself but as a function of measurement.

As a public figure, he actively ***worked for the peaceful use of atomic power*** and opposed Germany's developing nuclear weapons. He died on February 11, 1976.

Paul Dirac (1902-1984)

Whose passion was physics

Paul Dirac is considered one of the greatest theoretical physicists of this century. He wrote his first research paper in 1924 at the age of 22 and his last in 1979, producing in all 190 publications. Of these, 90 were pure research papers devoted to the development of the quantum theory.

Taciturn and introspective

Born on August 8, 1902 in Bristol, England, Paul Dirac was registered Swiss by birth; only in 1919 did he acquire British nationality. Young Dirac received his early education in the Merchant Venteurer's secondary school, which shared its buildings with the Engineering College of the Bristol University. As a result, Dirac even gained some practical experience.

About his early years, Dirac recalled in 1962, "I had no social life at all as a child... My father made the rule that I should talk to him in French. He thought it would be good for me to learn French that way. Since I found I could not express myself in French, it was better for me to stay silent than talk English. So I became very silent from an early age".

Thus he became taciturn and introspective. He could not revolt against his father's domination and compensated admirably by concentrating on mathematics with religious fervour, abetted by his father. Dirac did a lot of mathematics on his own, at a more advanced level than the rest of the class.

Graduated in electrical engineering

On completing school, he did not have any particular preference, science or engineering, for his profession. In 1918, Dirac entered the University of Bristol, as a student of electrical engineering, as he felt engineering would lead to a satisfying career. But he soon found that he was not good at practical work, so he did very little experimental work of any kind during his studies. ***He was disappointed that topics such as atomic physics*** (where he did original work later) ***and Maxwell's electromagnetic theory were not considered relevant to an engineering course.*** However, he came to know the value of the approximation theory, for he had till then believed that if one worked with approximations, there was an intolerable ugliness in the work.

Dirac recalled succintly, "If I had not this engineering training, I should not have had any success with the kind of work that I did later on, because it was really necessary to get away from the point of view that one should deal only with results which could be deduced from known exact laws, in which one had implicit faith".

Paul Dirac graduated in electrical engineering in 1921; but, for all his interests, he was unable to find a ***job as an engineer because of the depression in job market***. Luckily, he was offered free tuition by the Bristol University to study mathematics. Here he was introduced to the world of pure mathematical reasoning, which was very different in spirit from the engineering approach he met with earlier.

Career at Cambridge University

In particular, he came under the influence of Peter Fraser, a wonderful teacher, who was able to inspire young Dirac about basic ideas in mathematics. So he ***retraced his way from engineering to science.***

In the fall of 1923, Dirac enrolled at Cambridge University as a research student under Ralph Flower. As he met more people, he became less introverted. Within two years, he published seven papers (which became classics) and proved to be a talented physicist with a flair for fundamental research.

Dirac spent much of his time alone, in libraries and relaxed only on his solitary sunday walks. ***He now occupied himself with Heisenberg's***

revolutionary ideas about the study of atoms and tried to improve upon them. In this period of hectic research activity, he found time to submit his Ph.D thesis in May 1926.

Dirac's most enduring influence on physics was through the course of lectures he delivered for many years at Cambridge. Their content formed the basis of his celebrated work *The Principles of Quantum Mechanics,* which was published in 1930 and translated into other languages.

Dirac was elected Lucasian Professor in 1932 and held the chair at Cambridge University for 37 years till 1969, when he retired. Then he moved to the Florida State University in Tallahassee, U.S., where he remained loyal to physics, until his death on October 20, 1984.

Contributions

Among his many original contributions, the following are cited often:

- the quantum theory of the electron,
- the wave equation and
- the famous delta-function, which became one of the cornerstones in electrical engineering.

Honours

Dirac shunned honours and publicity. On the day his appointment to the Lucasian Chair was announced, he escaped to the zoo to avoid all the accolades. He had more than once refused honorary doctorates. In spite of this reservation, Dirac actually received all the honours a scientist could aspire for: Fellow of the Royal Society of London (1930), ***the Nobel Prize (1933)*** and Order of Merit (1973).

For Dirac, life was mostly science, and science, in his case, physics. His chief interest lay in fundamental research. His impact on the course of physics was through his research publications, his textbook on quantum mechanics and his lectures.

Advice to scientists

Dirac pursued the relationship between mathematics and physics at a far deeper level than is suggested by the principle of simplicity. He added the principle of mathematical beauty as the key concept in the relationship, which dominated all his work. So he gave scientists the following advice:

"The research worker, in his efforts to express the fundamental laws of nature in mathematical form, should strive for mathematical beauty. I should still take simplicity into consideration in a subordinate way to beauty." (*Dirac — A Scientific biography,* H. Krogh, Cambridge University Press 1990).

LORD RUTHERFORD (1871-1937)

THE FATHER OF NUCLEAR SCIENCE

"Could physics really be so simple as Rutherford disarmingly made it to be? Why was it all so simple ***afterwards*** and so complicated ***before***?" wrote one of the students recalling his undergraduate days at Cambridge in the early thirties (*Physics Today,* Dec. 1971).

Rutherford was not only one of Cambridge's illustrious professors but in physics, ***in the Cavendish Laboratory, he was the professor.*** He took personal interest in students' experimental work in his laboratory. One would receive occasionally a more or less unannounced visit from the professor at one's working bench. Rutherford would seat himself by his student and conduct quite a searching examination of the apparatus and method pursued. "What precisely are you doing?... Why?" He thus encouraged and admonished students, probed for the facts and always sought an explanation in simple terms. The laboratory produced men who later became legendary figures in science — C.T.R. Wilson, J. Chadwick, J.D. Cockcroft, M.L. Oliphant, J. Ratcliffe, F.S. Aston and H. Geiger.

Education and career

Rutherford was born in New Zealand on August 30, 1871. He had his early education in his native land and then worked as a research student in Cambridge University from 1895 to 1898. His working career was spread over three universities: McGill (1898-1907), Manchester (1907-1919) and Cambridge (1919-1937), succeeding J. J. Thomson (1856-1940) as Cavendish professor of Experimental Physics. Though Rutherford's name and his works are commonly identified with Cambridge and the Cavendish Laboratory, he was associated with other laboratories, too, at Montreal and Manchester. He died suddenly on October 19, 1937, at 66.

His discoveries

Rutherford is rightly called the father of nuclear science, just as Faraday is the father of electricity, on account of his fundamental contributions to the understanding of the disintegration and transmutation of radioactive elements. ***His emphasis was always on simplicity — simple idea and simple apparatus,*** but powerful conclusions resulting in a succession of brilliant discoveries.

- Discovery of the particles expelled from radium.
- Identification of the alpha and beta particles.
- Discovery of the neutron.
- Existence of the isotopes of hydrogen and helium.
- The nuclear theory of atomic structure.
- Atomic transmutation by bombarding nitrogen with alpha particles.

In 1919, Rutherford discovered that on collision with an alpha particle, an atom of nitrogen was converted into an atom of oxygen and an atom of hydrogen. Thus began ***the search that ultimately ushered in the age of nuclear energy.***

Rutherford was awarded the ***Nobel Prize in 1908***. It is interesting to recall his well-known pronouncement that "Chemistry is a branch of Physics and the rest of the sciences is stamp-collecting". The Nobel committee fittingly rewarded this observation by giving him the prize for chemistry!

Inspired lecturer

Besides his own personal contributions to science, Rutherford took to teaching seriously on all levels. He lectured officially thrice a week — monday, wednesday and friday at 12.00 noon in the Maxwell lecture theater of the Cavendish Laboratory — on 'The constitution of matter'. Rutherford's lectures were lively and up-to-date. His delivery style was characteristic: he always had a few loose sheets of notes in the inside pocket of his coat, which he pulled out at the beginning of each lecture, but referred to only occasionally.

Students were inspired listening to a great man relating an epic, which sounded rather like the story of some scientific expedition as narrated by its leader. They recalled vividly later — not so much what Rutherford thought about this or that, but rather how nature did its work and how a particular result had been discovered.

PETER KAPITZA (1894-1984)

BEACON OF CAMBRIDGE AND MOSCOW

Kapitza met Lord Rutherford and asked if he could return to work in the Cavendish Laboratory for a few months. As Rutherford's reply was discouraging, Kapitza unhesitatingly enquired about the accuracy Rutherford aimed for in his experiments. The latter was taken aback and replied, "2 to 3%". Kapitza pointed out that with 30 researchers working, one more would hardly matter, as he would come within the experimental error. Rutherford was thus persuaded to take him.

Early life

Kaptiza was born on July 8, 1894 in Kornstadt near Petrograd (Russia). A lively youth, he did not make adequate progress at the classical gymnasium and was transferred to another school, which was more scientifically oriented. He passed in 1912 with honours and entered the Petrograd Polytechnical Institute, from where he graduated in 1919. Soon after, he was appointed to a teaching post in this institute, which he held till he left for England in 1921.

Visit that changed the career

Kapitza was a legendary figure in Cambridge between 1921 and 1934. He along with A.E. Joffe, arrived in Cambridge as a member of a Commission of the Russian Academy of Sciences. The visit changed the course of his career.

Though Kapitza planned to stay only for the winter of 1921, he remained in England till 1934, a span of 13 years. On Rutherford's advice, he took up the study of the velocity of alpha particles by measuring the curvature of their tracks in a magnetic field. After some early failures in the production of strong impulsive fields, he hit upon a new method — a large dynamo which, on short-circuit through a suitable coil could generate high power for a very short time and so produce a field of a million gauss.

In 1930, Kapitza suggested to Rutherford regarding setting up a new laboratory, which could house not only the high-field equipment but also provide cryogenic facilities for advancing towards lower temperature. Actually, this was the main motivation for his later work on hydrogen and helium liquefaction. Rutherford supported this idea and the Mond Laboratory was born in 1933 and Kapitza, its first director.

Meanwhile, his position in Cambridge was rapidly being consolidated. In January 1923, he was officially ***registered for the Ph.D. degree, with retrospective effect from October 1921,*** which made possible the award of the degree in June 1923. In August 1923, he got the prestigious Clerk Maxwell scholarship, which improved his finances. He was the first Russian to be elected Fellow of the Trinity College (1925) and the first foreigner in 200 years to become an FRS (1929).

Style of his lectures

His other academic work was supervision of a few research students — P.M.S. Blackett, J.D. Cockcroft, D. Shoenberg — and some teaching activity. Students found his lectures fascinating, not only for their content but also for the intriguing presentation.

The weekly seminar, which he started in October 1922, came to be called the Kapitza Club. It played a notable role in stimulating physics in Cambridge. Thus, scientifically he had grown strong and mature. Kapitza was impressed with the English commitment to individualism in scientific research and to insistence on obtaining results. He wished that the Soviet would follow this pattern.

From 1926, Kapitza visited his country every summer. In August 1934, he left for what was intended to be a short trip to attend a Congress in Moscow in honour of Mendeleev (1834-1907). But it turned out to be a permanent return. For a time he sulked and contemplated taking up biophysics with Pavlov (1849-1936), if he could not continue his Cambridge work.

Rutherford's assistance for a new institute

However, by the end of the year, he was asked to plan a new institute at Moscow, the Institute for Physical Problems (IPP). Only later did he learn that he had been appointed director! The Institute was in part almost a Chinese version of the Mond Laboratory, with the same themes for work — magnetism and low-temperature physics.

Rutherford exerted his influence and helped Kapitza. He arranged with the Cambridge University to transfer for a payment of £30,000 nearly all the Mond equipment and duplicates of some items along with Kapitza's two assistants for setting up all the equipment in the IPP. The institute started functioning by December 1936.

Kapitza took up two new lines of research, which became his main preoccupation for the next ten years. One of these, the study of the transport properties of helium-11, led to a series of remarkable discoveries and the award of ***two Nobel Prizes, one to Landau*** (1908-1968) for his theory of quantum fluids and the other, very belatedly in 1978, ***to Kapitza himself.*** The second was the development of a new and more efficient method of liquefying air.

Honours

Kapitza carried on his scientific work in IPP till his death on April 8, 1984. His work was interrupted twice: from 1941 to 1943, when the institute was evacuated to Kazan and from 1946 to 1954, when he was under house arrest at Nicholas Hill for refusing to work on the nuclear weapon programme.

One of his main achievements at Kazan was the rapid development of an industrial-scale plant for oxygen production using the expansion-turbine method. This was in spite of lack of funds and extreme shortage of food, as the Soviet Union entered the war. ***Kapitza got all the coveted awards*** — Stalin Prizes in 1941 and 1943, Orders of Lenin in 1943 and 1944, the title of Hero of Socialist Labour in 1945, ***Nobel Prize in 1978.***

In his country house at Nicholas Hill, he managed to carry on his scientific work aided by his two sons, until he was reinstated as Director in 1954. Kapitza then returned to Moscow and continued with his work on very hot glowing plasma. He was made chief of the Sputnik programme (1955). ***In the style of running his institute, Kapitza modelled himself closely on his revered professor.*** Like Rutherford, he kept in touch with what everyone was doing and making shrewd comments, even if the topic was remote from his personal research.

A daring act

Kapitza gave unstinted loyalty to his colleagues. When Landau, Professor of physics in his institute, was arrested in 1938, Kapitza showed great courage by ***taking up the matter with Stalin, an act aptly described as "jumping naked into a den of lions".*** He got Landau released under his personal guarantee.

The bonds that united Kapitza and Rutherford in their scientific quest were strong and deep. Kapitza had great affection for him and openly showed his admiration for Rutherford's genius. He discovered a fanciful nickname 'crocodile' for his professor, more out of awe and admiration than in any spirit of light-hearted teasing. In turn, Rutherford seemed to have been captivated by Kapitza's boldness in scientific vision and in treating Rutherford with much less reverence than the man was accustomed to.

Peter Kapitza is an example of a scientist who, at a personal level, established technical cooperation between Russia and England, and produced results of fundamental value in physics and engineering. He achieved the rare distinction of being both an FRS (while being a Soviet citizen) and a member of the Soviet Academy of Sciences.

Rutherford stamp in Russia

Kapitza was grateful for the encouragement and fatherly care that he received from Rutherford. He organised a ***Rutherford Centenary Colloquium in Moscow*** and got the Soviet post office to issue a Rutherford commemorative stamp in 1971, which the British post office was unable to do.

In a letter to Niels Bohr, Kapitza expressed succinctly what he owed to his mentor: "I ***learned a great deal from Rutherford,*** not physics but ***how to***

do physics. Rutherford was not a critic; you, probably, like me never heard him arguing, neither in science nor in questions of life. But his influence on everybody around him was felt by his example and by his opinion, always brief and in the long run, correct".

(*Kapitza in Cambridge and Moscow,* Boag (Ed.), Rubinin and Shoenberg, North-Holland, 1990)

James Chadwick (1891-1974)

Nuclear scientist by accident

Chadwick proved in 1932 the existence of the neutron, which marked the beginning of the modern era in the understanding of the atomic nucleus. It led to ***a new tool in atomic disintegration,*** capable of penetrating and splitting the nuclei of even the heaviest elements.

Early education

James Chadwick was born on October 20, 1891 in Cheshire, England in a working-class family. His father was a laundry proprietor in Manchester. After attending the local high school, he entered in 1908 the Manchester University. He graduated (1911) with an honours degree in Physics and spent the next two years under Professor Rutherford. He investigated various radioactivity problems, for which he obtained the M.Sc degree in 1913.

The same year, he was awarded the coveted 1851 Exhibition Scholarship for doing research work at the Physics Institute, Charlothenburg under Professor Geiger (1882-1945). He was the first to obtain (1914) the energy spectrum of the beta particles from radioactive substances; he posed the problem that led Wolfgang Pauli (1900-1958) to postulate the existence of the neutrino in 1931.

A trivial mistake that changed the career

During World War I, he was interned in Ruhleben, Germany. He could return home only in 1919. It is said ***he became a physicist by attending the wrong interview with Rutherford*** at the University of Manchester. Being too shy to own his mistake, he joined Rutherford in making studies of the properties and structure of atomic nucleus. And when Rutherford moved to the Cavendish Laboratory, Cambridge, Chadwick was the only physicist invited to accompany Rutherford.

Rutherford quickly found in Chadwick a superb talent for research and entrusted him with the work on artificial nuclear transformation. It turned out, ***he thus set the agenda for a string of Nobel Prizes!*** He had succeeded in 1919 in disintegrated atoms by bombarding nitrogen with alpha particles with the emission of a proton.

Discovery of neutron

Rutherford has speculated about the existence of a neutral proton: Chadwick seized the opportunity to search for the neutron — an elementary particle devoid of any electrical charge. Using powerful alpha-emitters from an American cancer hospital, he found that the radiation generated by alpha particles hitting beryllium could eject particles from a whole catalogue of elements, which he reported in a lucid letter for *Nature.* Chadwick proved in 1932, the existence of the neutron, which marked the beginning of the modern era in the understanding of the atomic nucleus. It led to ***a new tool in atomic disintegration, capable of penetrating and splitting the nuclei of even the heaviest elements.***

Break with Lord Rutherford

For this epoch-making discovery, he was awarded the Nobel Prize for physics in 1935. He intended to use the cash prize in part to pay for a cyclotron, which would replace radioactive sources for nuclear experiments Here the break set in their cordial relationship, as Rutherford always regarded Chadwick as his heir apparent. Lawrence Bragg (1890-1971) became in 1937 the director of the Cavendish Laboratory, although Chadwick felt he should have been asked. On being elected to the Lyon Jones chair of Physics in the University of Liverpool, Chandwick shifted in 1935 to his wife's home city, where he had plans to build a cyclotron. Chadwick missed heading the world-renowned Cavendish Laboratory.

Physicists in the University of Copenhagen bombarded uranium-235 with neutrons, which was accompanied by a large release of energy. By June 1940, Chadwick was coordinating research in several universities to try whether a weapon was practicable.

A year later, on behalf of the committee code-named MAUD, he set out in his clear style the explosive power of a uranium-235 bomb and the method to produce the bomb. The U.S. accepted the MAUD report and Chadwick's verdict that an atom bomb was feasible. General Leslie Groves was put in charge of the project and now, everything depended in winning the confidence of the dictatorial Groves who wanted to exclude foreigners.

Viceroy in General Groves' team

How the gloomy Chadwick pulled off this feat is described in a recent biography. "He first enchanted the top American physicists by daring to argue with the general and then Groves chortle by thrashing the whiz kid Robert Oppenhemier in a highly technical argument. Groves saw that the plain-dealing, respected physicist was just the man to be his viceroy for his team" *(The Neutron* and *the Bomb:* A *biography of Sir James Chadwick,* Andrew Brown, Oxford University Press, 1996).

By the time Chadwick returned home in 1946, he was a worn-out person: too many years of seven-day working schedule, interrupted by illness, ruined his health. He retired in 1943 from his active physics career and found sanctuary as Master of Caius College, Cambridge. He never visited the Cavendish Laboratory.

Honours

Chadwick received several academic honours: Copley Medal (1950), Franklin Medal (1951), honorary doctorate degrees from the Universities of Birmingham, Dublin, Edinburgh, Liverpool, Oxford and Montreal and membership of several foreign academies. Chadwick was knighted in 1945. His lifelong shyness had become pathological: ***when the King of Sweden handed the Nobel Prize to him,*** he fumbled and a bystander had to come to his help to pick up the cheque from the stage.

He was a member of the U.K. Atomic Energy Authority (1957 to 1962). He ensured a major role for his country in CERN, the European laboratory for Particle Physics created in 1952. He died on July 24, 1974, at Cambridge.

Christian Doppler (1805-1853)

Celebrity in Physics by a Single Discovery

> "The principle was put forward by Doppler with the sole object of explaining the colour of stars, but the deductions from observational evidence which he marshalled to support it were completely erroneous... Still, in recent times it has risen to the position of providing valid evidence for an expanding universe. The modest Austrian would be astonished to see the scope and the importance of ***the work which in his lifetime aroused but little attention"***
> — E.N. da Andrade.

Self-educated

Doppler, the son of a noted stone mason was born on November 29, 1805, in Salzburg, Austria. He gave early evidence of mathematical ability, which was recognised by the mathematician Stampfer. On the latter's advice, he studied in the Polytechnic Institute at Vienna from 1822 to 1825. Not impressed with the curriculum, he returned to Salzburg ***to pursue his studies privately.*** Doppler completed the course in the gymnasium in an unusually short time, while giving tuition privately in mathematics and physics.

His career

From 1829 to 1833 he was employed as mathematics teacher in Vienna. He had completed preparations for emigrating to the U.S. in 1835, when he obtained a position in the State Secondary School in Prague. In 1841, he became professor of elementary mathematics at the State Technical Academy, while holding this post he published his celebrated principle.

Doppler moved in 1841 to the Mining Academy at Schemnitz. He was beginning to be recognised as a man of distinction and was awarded (1848) an honorary doctorate by the University of Prague. A little later, he returned to his country and was made the ***professor of experimental physics at the University of Vienna, which was the goal of his life.*** He did not long enjoy this post of academic distinction, for he died of lung trouble on March 17, 1853.

On the centenary of his birth, various posthmous honours in Prague, Salzburg and Vienna sprang up: streets were named after him and a bust of him was erected in 1901 in the University of Vienna.

Doppler principle

The paper in which Doppler established the principle by which his name lives was published in 1843 under the title *'On the coloured light of the double stars',* at that time there were still physicists who were not convinced of the validity of the wave theory.

Colour depends on the frequency of the waves stimulating the eye and anything that changes the interval between successive impulses changes the colour perceived. As long as the source and the observer are at rest, the frequency observed is the frequency of the light emitted by the source; but if the observer is moving towards the source, the frequency must increase: if he is moving away from it, the frequency decreases. ***Doppler mentioned in his article the application of his result both to acoustics and optics,*** with the remark that, at first hearing, this might seem purely academic.

Unique experimental verifications

The acoustical effect was verified by Buys Ballot in 1845, despite the necessary scientific instrumentation not being then available. He used a locomotive to draw an open carriage with several trumpeters. ***The observers were trained musicians,*** who estimated the changes of pitch by ear. He repeated the experiments with observers in the car and trumpeters by the side of the track. Despite the locomotive noise vitiating the observations, the results agreed reasonably.

H.C. Vogel improved in 1876 the accuracy of the tests, using a locomotive with the whistle of pure tone. ***The notes of the approaching and receding whistle were matched on a violin.*** K.R. Konig devised (1865) a single demonstration, which is of interest as illustrating the method for extending the application of the effect with radio waves.

Buys Ballot criticised Doppler's assertion that the eye could detect the differences of colour due to the movement of stars, to which Doppler replied in a rather assertive fashion.

It was Fizeau (1819-1916) who answered (1848) the above criticism, by pointing out that the apparent colour of the source would be changed at high relative velocities by the motion; that is, an object moving towards the observer would appear bluer and one moving away would appear redder. He further devised an ingenious experiment to demonstrate the effect. Still, the validity of the general principle did not gain the acceptance of Doppler's contemporaries, the chief antagonist being Joseph Petzval (celebrated, for his work on lenses) whose opposition began in 1852 and continued for some years.

Practical applications in different fields

But with light, laboratory demonstration is much more difficult than with sound. This was achieved by Belopolski (1900), Fabny and Buisson (1919). The applications to astronomy were published in a paper by Armand Fizeau (1848) and William Huggins (1868). The Doppler shift in the spectra of celestial objects was used to measure the rate of recession or approach relative to the earth.

Other applications are in completely different fields — in radar, aerial navigation for determining aircraft velocity relative to the earth, in remote sensing, tracking artificial satellites circling the earth; and in biomedical instrumentation. Modified by the theory of relativity, ***the Doppler principle has now become a major astronomical tool.***

Like Argand (1768-1822), who became a celebrity in mathematics by a single invention, Christian Doppler is a name known only in connection with the effect named after him. In the case of other men of science like

Coulomb (1736-1806), Faraday (1791-1867), Gauss (1777-1855) and Mach (1838-1916), the laws that bear their name represent a small part of their contributions. But, Doppler contributed nothing else to science that deserves to be remembered. Today, his finding has wide applications from detecting speeding vehicles to high-energy physics.

Extension by Stark

Another epoch-making application was demonstrated by John Stark (1905), when he obtained the effect from a shift of spectral lines emitted by a beam of fast ions (canal rays) in a gas discharge tube operated at high voltage. This work gained importance for the proof it gave to the electrical state of the atoms emitting spectral lines and of the molecules emitting band spectra. ***In the Nobel citation, when Stark was awarded the prize in 1919, the main reference was to the Doppler effect,*** which he employed in his investigations, the Stark effect receiving supplementary mention.

William Roentgen (1845-1923)

The first physics Nobel Laureate

It was a remarkable day. On December 22, 1895, Roentgen brought his wife to the laboratory. He observed in a photographic plate the image which showed the shadow thrown by the bones of her hand, including that of a ring she was wearing. This was the first X-ray photo.

Early years

William Roentgen was born in a family of merchants on March 27, 1845 at Lernep, Germany. In school, he was not a studious boy, preferring outdoor life in the country to classes and showing marked aptitude for the construction of mechanical contrivances. In 1862, he entered the Ultrecht Technical School. He could not complete his studies as a regular student, as he was expelled for refusing to identify a classmate who had made caricatures of one of the teachers.

In 1865, Roentgen passed privately the entrance examination to the ETH at Zurich (Swiss Federal Institute of Technology) and began his studies in mechanical engineering. Roentgen attended the lectures of the well-known physicists Clausius (1822-88) and Kundt (1839-94), who exerted great influence on his development. He received his diploma in 1868 and doctorate degree for his thesis on 'States of gases' in the following year.

Professor at Wuerzberg, Geissen and Munich

In Zurich, Roentgen became assistant to Professor Kundt and went with him (1869) to the University of Wuerzberg. The University refused to appoint him to any academic position, since he lacked the academic credentials ('habilitation' requirement). So he moved again with Kundt to University of Strasburg (1876-78). The series of papers he produced won him the chair of physics at Geissen (1879-88). Here he built such a reputation that the same University of Wuerzberg ***invited him to take up the joint posts of professor and director*** of the Physical Science Institute.

As Kundt's assistant at Wuerzberg, he formed the life-long habit of building his own apparatus. Being a conscientious experimenter, he preferred to work alone in the laboratory. He was thus able to develop acute powers of observation, to detect and measure extremely small effects such as the compressibility of liquids and solids, the rotation of plane of polarisation of light in gases.

He found among his colleagues, Helmholtz (1821-1894), and Lorentz (1853-1928). In 1900, he accepted the chair of Physics at University of Munich from which he retired in 1920.

Roentgen returned to his country house at Weihelm near Munich, spending his time in his private library and enjoying long country walks. He died on February 10, 1923.

Fundamental work

Among the problems that Roentgen studied, the following embrace fundamental areas:

- electrical characteristics of quartz,
- effect of pressure on the refractive index of fluids,
- compressibility of fluids,
- spreading of oil drops on water and
- modification of the planes of polarisation by electromagnetic fields.

However, ***Roentgen's fame rests on two pieces of work***, which would be seen to lie far outside his fieldwork.

i. The first is the magnetic effect produced in a dielectric, such as a glass plate, when it is moved between two electrically charged condenser plates. He showed the application of Maxwell's theory to explain the effect quantitatively. This led to the theories of Lorentz and ***heralded the age of modern physics.***

ii. While experimenting with the flow of electric current in a discharge tube (now known as Crookes tube), Roentgen observed that a nearby piece of barium platinocyanide gave off light. He covered the discharge tube with a black card and still noticed that a fluorescent screen lying some distance away was glowing brightly. It was Roentgen who theorised that these are new and different kind of rays emitted by the glass wall of the tube that was opposite the cathode. ***He thought the rays were unrelated to light so he called them X-rays.*** Later Max von Laue (1879-1960) showed that they were part of the electromagnetic spectrum but of very short wavelengths (0.01 to 50 angstroms).

It was ***a memorable day. On, December 22, 1895,*** Roentgen brought his wife to the laboratory where he immoblised her hand in the path of the rays. He observed in a photographic plate the image, which showed the shadow thrown by the bones of her hand, including that of a ring she was wearing. This was ***the first X-ray photo*** the world has known.

Honours

Prizes, medals, membership of several learned societies were showered on him. The accolades for this modest scientist were the erection of his statue in Berlin and the award of the first Nobel Prize for Physics in 1901. ***He donated the prize money to scientific work*** at the University of Wuerzberg.

Roentgen's reticence caused him to shun public engagements, so he declined to give the expected Nobel lecture. But in his address (1894) as Rector of Wuerzberg University, he gave a clear mandate to men who rule the destiny of universities:

"The University is a nursery of scientific research and mental education, a place for the cultivation of ideals for students as well as for teachers... for this reason, one should endeavour in filling vacant places, to ***choose men who have distinguished themselves as investigators and promoters of science*** and not only as teachers; such a one, who takes his task seriously, fundamentally follow purely *ideal goals".*

C.V. Raman (1888-1970)

India's first Nobel Laureate in physics

Raman's first trip abroad was to the U.K. to attend the congress of Universities held at Oxford (1921). He met the leading scientists and delivered lectures on his research in optics and acoustics. The return trip was historic as it was during this voyage that he became fascinated with the deep blue colour of the Mediterranean.

Educated at Vishakapatnam and Madras

Chandrasekara Venkatraman (known as C.V. Raman) was born on November 7, 1888 in Trichonopoly (now Trichy), Tamil Nadu, South India in an erudite family. His father was a teacher in a local high school, from where he shifted in 1892 to a college in Vishakapatnam (Andhra Pradesh) as lecturer in physics.

Raman spent the next ten years in Vizag and joined in January 1903 the Presidency College, Madras, for his undergraduate studies. He achieved the highest distinctions in B.A. (1904) and M.A. (1907) examinations, winning the first place in the Presidency. Besides a flair for physics, he developed a liking for the English language, awakened by the classes of Professor Bilderberk.

Raman enjoyed the measure of academic freedom given by the Professor of physics R.L. Jones and derived the greatest benefit from it. Only equipment needed for routine laboratory work such as prisms, gratings, lenses, sonometers were available. With these, he started doing simple experiments on his own.

Research publications as a student

For example, the diffraction of light by a rectangular slit was treated in standard textbooks: but Raman took up ***the problem of unsymmetrical diffraction.*** He wrote down his experimental findings and sent the paper (1906) directly to the prestigious journal *The Philosophical Magazine* in London under the title 'Unsymmetrical diffraction bands due to a rectangular aperture'. This was followed in 1907 by a second paper 'The curvature method of determining the surface tension of liquids'.

A research paper by Lord Rayleigh on the subject of sensitiveness of the ear to pitch inspired Raman. The two exchanged correspondence, in which Lord Rayleigh addressed Raman as professor, whereas Raman was still a student!

Raman also became interested in the vibrations of a stretched string. It occurred to him that, by suitably ***altering Melde's experiment it would be possible to obtain results surpassing in beauty*** those found by Melde himself. All these ideas were sown in his mind, while still a student in the Presidency College.

Palit Professor (1917-1933)

A career in scientific research being at that time almost nonexistent in the country, Raman entered in 1907 the Finance Department of Government of India though a competitive examination. He retained this position for ten years, mostly in Calcutta but with brief periods on transfer to Rangoon and Nagpur.

But Raman's passion was for physics. How did he find the opportunity to realise the dream of continuing his scientific work? This was provided by two great citizens of Calcutta — Dr. Mahendra Lal Sircar, founder of the Indian Association for the Cultivation of Science (IACS) and Sir Ashutosh Mookerjee, the Vice-Chancellor of Calcutta University.

Raman described the latter's offer of the post of Palit Professor as an act of courage. But for this action, he acknowledged, "My scientific career would long ago have suffered an abrupt termination." He occupied the chair from 1917 to 1933. Besides, from 1919 he acted as the secretary of the ICAS and brought to Calcutta many aspiring young men to undertake research in optics and acoustics. He described this Calcutta period as the 'golden era' of his career.

Raman devoted his leisure time to ***experiments on the acoustics of Indian musical instruments*** at the association. This resulted in thirty papers being published in prestigious journal such as *Nature, Philosophical Magazine* and *Physical Review.* His name was now familiar to scientists in Europe and America.

Blue of the Mediterranean Sea

Raman's first trip abroad was to the U.K. to attend the Congress of Universities held at Oxford (1921). He met the leading scientists and delivered lectures on his research in optics and acoustics. The return trip was historic as it was during this voyage that he became fascinated with the deep blue colour of the Mediterranean Sea. Lord Rayleigh (1842-1919) had some years earlier arrived at the conclusion that this was due to the reflection of the sky blue by the ocean's surface. Raman used the simplest equipment, just a Nicol prism, to examine the colour of the ocean. His fertile imagination now led him to ***eliminate the skylight reflected by the sea surface in the crossed position of Nicol prism. He found the colour of the sea "a much richer blue than that of the sky".*** He thus gave ample proof that Lord Rayleigh's earlier conclusion was not correct.

Raman effect

This led to a serious study of the scattering of light, culminating in the discovery of the Raman effect, early in February 1928. "When a beam of monochromic light is scattered by a transparent liquid, the scattered light contains frequency of very low intensity not present in the incident radiation." He outlined his experimental results in his first book *Molecular Diffraction of Light* (1922).

During 1921 to 1928, Raman attracted a group of dedicated research workers who included K.R. Ramanathan, K.S. Krishnan, Ashutosh Dey, S. Bhagavantam and a few others. There was a close collaboration between Raman and Krishnan in the Raman effect.

Creating a centre of excellence at Bangalore

Shifting to Bangalore in April 1933, Raman was invited to become the first Indian Director of the Indian Institute of Science, Bangalore, founded in 1911 by the munificence of J.N. Tata. He fell in love with Bangalore so much that he made it his permanent home. ***Raman saw that the Institute was rich in prestige but short of academic distinction,*** as it had not produced any outstanding scientific discoveries.

An opportunity for inducting new talent was presenting itself, since many eminent scientists were fleeing Hitler's Germany. Raman felt that by inviting some of these outstanding men of science to serve the Institute, he could make it a centre of excellence. He got Max Born (1882-1970) appointed as a professor and made an offer to Schrodinger (1887-1961), the founder of wave mechanics. By these actions, he was confident ***he could create a real scientific movement.*** All these proved to be miscalculations: his strong personality and dynamic management style forced his resignation from the post of Director. "What a great setback", the biographer G. Venkatraman ruefully observes, "it was for our science when one recalls the advantage gained by America from immigrant scientists!" *(Journey Unto Light,* published by the Indian Academy of Sciences, Bangalore, 1988)

Despite this, Raman continued as Professor of physics till his retirement in July 1948 and turned the department into an active centre of research in diverse fields, X-ray scattering, physics of the diamond and lattice dynamics. Homi Bhabha (1909-1966) and Vikram Sarabhai (1919-1917) worked for some time with Raman.

Raman Research Institute

In 1949, Raman moved to his own institute, which he built on a lovely piece of land gifted by the Maharajah of Mysore. He equipped the institute with beautiful museums, lecture hall, laboratories, and accommodation for staff and bequeathed all his personal wealth to this centre. Raman spent the last ten-years of his life (till November 21, 1970) in investigating the perception of light and colour, all by himself at an advanced age. He published his second book *The Physiology of Vision* (1968). The Raman Research Institute has gained international stature and is bustling with research activity in liquid crystals, astrophysics and radio astronomy.

He founded the Indian Academy of Sciences at Bangalore in 1934. He was the founder editor of its proceedings and of the *Indian Journal of Physics.*

Gandhi memorial lecture

Next to his love for roses, Raman liked giving lectures. He was a lecturer par excellence: a stentorian voice, a graphic delivery producing a powerful impact on the audience, a knack to reduce any scientific topic to the simplest terms. Every year on October 2, he would give the Gandhi Memorial Lecture, where he used to select a scientific topic and make a masterly presentation.

Honours

Raman emerged as a true international figure: FRS (1924), Knight (1929), Hughes Medal (1930), ***Nobel Prize in Physics (1930)*** and Franklin Medal (1942). He was venerated as the Father of Indian Science. The Government of India appointed him the first National Professor in 1948 and decorated him with the ***Bharat Ratna in 1954***.

Enrico Fermi (1901-1954)

Who embraced all physics

In Enrico Fermi, the theoretical and experimental physicists were combined in a supremely complementary and creative way. This is an unusual feat in the twentieth century, when increasing specialisation had tended to narrow down the field of study.

Self-taught

Fermi was born on September 29, 1901 in Rome, Italy. He received traditional education in the public schools of Rome, but he was ***self-taught, dependent on the secondhand books that he bought in local shops.*** His marvellous aptitude for mathematics and physics manifested itself very early. He said later that, by the age of ten, he had succeeded in understanding practically unaided why the well-known equation $x^2 + y^2 = r^2$ represents a circle.

His science formation

A colleague of Fermi's father, Adolfo Amidei, recognised the boy's extraordinary ability and lent him books on mathematics and physics, selected and marked in a pedagogically graded fashion. Thus by 17, he had acquired a thorough foundation of classical physics, comparable to that of a graduate student in an university.

Fermi mentioned, later in life, having studied in private the works of Poisson (1781-1840), Poincare (1854-1912), Planck (1858-1947) and Richardson

(1868-1928). In particular, the two volume treatise on mathematical physics by A. Caraffa, a Jesuit, prepared his science foundation.

In the grade school, Fermi was the sort of ***model student who succeeded in everything.*** He struck a friendship with Enrico Perisco; their common scientific interests blossomed into a lifelong friendship and they went on to become the first professors of theoretical physics in Italy.

Forecast by an examiner

On Amidei's advice, Fermi competed for the entrance examination for admission to the 'Scuola Normale Superiore' at the University of Pisa, where his education would not cost any expense to his family. His essay for the competition, 'Distinctive properties of sound' began with the enunciation of the partial differential equation for vibrating reeds and concluded with his original solution using the Fourier series. The examiner, a good mathematician in his own right, was thrilled to receive such ***an original essay from a boy just out of high school.*** He was convinced of the candidate's genius and conveyed to Fermi his evaluation, which Fermi remembered for many years to come.

Fermi received his doctorate from the University of Pisa in 1922. To acquire direct interaction in modern physics, he went to Gottingen in 1923 to work under Max Born (1883-1970) and moved to Leyden in 1924 to work under Paul Ehrenfest (1850-1953). Fermi then returned to his native country and served as a lecturer in the University of Florence, teaching mechanics and mathematical physics.

Fermi level

The Florentine period saw his first major contribution to physics, the discovery of statistical laws, which govern particles subject to Pauli's exclusion principle or 'fermions', as they are known today. His discovery led to an understanding of many fundamental questions such as how electrons in metals conduct (Fermi level), why electrons do not contribute to the specific heat of substances.

Working for the rebirth of Italian physics

In 1926, Fermi was offered the Chair of Theoretical Physics at the University of Rome, where he remained till 1938. It was his good fortune that he met Senator Morio Corbino, Professor of experimental physics. He immediately recognised that Fermi was a prodigy and became his patron. He saw Fermi as the instrument for one of his fondest aspirations — the rebirth of Italian physics. Corbino did not miss any occasion to get recognition of Fermi's work. In 1929 Fermi was appointed by Mussolini to the newly created Italian Academy at Rome and was the only physicist so honoured at an unusually young age of 28.

His students' perception

The period 1926-38 was Fermi's golden age. His knowledge and interests embraced all physics. He preferred concrete problems to distrusted theories; any specific problem in any field of physics — classical mechanics, spectroscopy, thermodynamics, solid state theory — fascinated him and was a challenge to his ingenuity. His students saw, while just talking to him, a simple and clear explanation develop that resolved some puzzling phenomenon. They described Fermi as "a steam roller that moved slowly, but knew no obstacles".

Fermi made little use of books: Laska's collection of mathematical formulae and Landolt's tables of physical constants were the only reference books he had. ***If he needed formulae, he would derive it faster than one could find it in a book!***

Migration to the U.S.

The Fascist racial laws, which directly affected his wife and deeply offended his sense of fairness, induced him to emigrate to the U.S. After receiving the ***Nobel Prize in December 1938,*** he did not return to Italy but directly proceeded to New York, where he accepted a professorship at Columbia University and began ***the second phase of his career.***

At Columbia, the study of neutrons has been one of the major subjects of investigation. Without the announcement of the discovery of fission and without the outbreak of the world war, it is probable that Fermi would have left neutron physics and turned to another subject. His intuitive feeling for the behaviour of neutrons was like that of a radio engineer for the behaviour of circuits. He did not need to calculate in order to predict the results of an experiment; but, true to his nature, after each experiment, he performed detailed calculations, which he recorded in an extremely orderly fashion.

Nuclear chain reactions

During World War II, he worked on the development of nuclear energy on a large scale resulting in nuclear chain reactions at Chicago and Los Alamos. At the end of the war, he moved on to the University of Chicago where he served until his early and unexpected death on November 29, 1954.

"Fermi gave to science all he had and with him disappeared the last universal physicist in the tradition of the great men of nineteenth century, when it was still possible for a single person to reach the highest summits both in theory and experiment, and to dominate all fields of physics".

(*Introduction to Collected Papers of E. Fermi,* Emilio Segre, The University of Chicago Press, 1962)

Laplace

SECTION THREE

Niels Henrik Abel

MATHEMATICS

Gauss

RENE DESCARTES (1596-1650)

FOUNDER OF ANALYTICAL GEOMETRY

Out of his development of the Cartesian graph, a fundamental idea blossomed giving impetus to the concept of variables and functions. It gave birth to the branch, now known as analysis.

Studies in a Jesuit College

Rene Descartes (pronounced dakaert) was born on March 31, 1596, in La Haye (France). His father was a member of the nobility. He lost his mother a year after his birth. He was raised by his maternal grandmother and by a nurse to whom he was devoted.

In 1606, Descartes entered a Jesuit College that was established for the education of the young nobility and completed in 1614, the eight-year course of studies. His studies left him with a feeling of dissatisfaction at the extent of his own ignorance.

But, Descartes left behind a mark of extreme brilliance. He spent the period from 1618 to 1626, divided between participation in many campaigns of the Thirty Years War in Dutch, Bavarian and Austrian armies, excessive travelling and amusement, and then complete retirement from 1626 for the next 20 years. He never married.

While doing military service without salary, Descartes spent most of his time in the study of mathematics. By the age of 22, he began to develop his analytical geometry under the title *'La Geometry'* by which the mathematicians still affectionately refer to it. ***It set forth an idea that a pair of numbers can determine a position on a surface:*** one number x as a distance measured horizontally, the other y as a distance measured vertically (Cartesian coordinates).

In the concept of coordinates, Descartes gave a method commented to be ***"the greatest single step ever made in the progress of exact sciences".***

Descartes started, in 1619, work on his magnum opus *Discourse on the method of rightly conducting the reason,* which stamped him immediately as one of the great thinkers of his age. He postulated his notion of the unity of all sciences, which was a ***revolutionary concept that contradicted the notion of Aristotle.***

Analytical geometry as a footnote

The book contained three minor appendices on the behaviour of lenses, meteors and geometry. A 106-page footnote outlined analytical geometry, which he did not pursue further.

In the next three centuries, this was to outstrip other branches in creating the science Descartes had dreamt of; ***graph paper, with its use of crossed lines, had not yet been invented.*** Descartes evolved a theory for 'conic sections' on which Apollonius had lavished so much ingenuity.

The graphing technique, he employed point by point, for sin x created a curve on paper: it is an endless wavy line, the exact graphic equivalent of the pulsating ebb and flow of electric current in a power cable, which has practical applications in electrical engineering.

Descartes showed, for instance, that all quadratic equations when graphed as connected points become straight lines, circles, ellipses, parabolas, or hyperbolas. Going beyond the quadratic, he established that each class of equations of higher degree brings into being a whole new tribe — figure of eight, hearts, hummocks and petal shapes. ***Analytical geometry thus transformed every equation into a geometric shape and every geometric shape into an equation***. Branches of mathematical thought such as trigonometry and logarithms were now brought into the main stream.

Out of his development of the Cartesian graph, ***a fundamental idea blossomed, giving impetus to the concept of variables and functions***. It gave birth to the branch, now known as 'analysis', which encompasses most of the higher mathematics.

In all manner of practical applications, all the inter-relationships and fluctuations measured can be converted into sets of numbers. For example, a bridge designer can write the curve of a slack cable as an equation and through its *x–y* coordinates, gain more understanding of the stresses on the bridge.

Descartes applied Snell's law of refraction for the first time to the course of the rays in the rainbow. The colours were, however, left unexplained, and this success remained for Newton.

Moving to a village

Ironically, the man who thus opened the gateway to higher mathematics did not pursue it. Instead, Descartes plunged into philosophy, with a determination to recreate it.

From 1626 to 1629, he resided in Paris but soon found that the pressures of social life kept him from his work. So he moved to a village near Amsterdam. He cherished the solitude of his life, never rising before 11 a.m. He was eternally seeking peace, fleeing from one rented villa to another. He studied and wrote enormously.

Revelation in an army camp

Here, his work saw the light of print in 1637. It was 18 years earlier that it revealed itself to him on November 10, 1619 in an army camp on the banks of the Danube. It was a cold day. Descartes spent it meditating in a small, heated room, known then as a stove. He formulated the doctrine that all knowledge should be worked out in terms of mathematical reasoning.

The more he looked for fundamental truths, the fewer he could find. In the end, he came out with the famous formula: ***"I think, therefore I am",*** by which he meant that the basis for understanding the real world starts from man's ability to use his own mind. ***This vision of Descartes has remained the ambition of modern science.***

A new methodology

Descartes' research in mathematics and physics led him to see the need for a new methodology or a new way of thinking. His other major work '*Rules for the Direction of Mind'* was written by 1629 and was in circulation in manuscript form; it was not published until 1701.

Descartes' findings can be summarised as follows: literature and eloquence persuade man, but they do not tell the truth. Mathematics does grasp the truth, but the certainty of its reasoning seemed to have only practical applications.

The revelations of religion seem as mysterious to the learned as to the ignorant. Philosophy had been studied by the best minds; and yet no single thing is to be found in it, which is not subject to dispute. Descartes even came to suspect that the solid foundations of science were so far from firm.

Invitation from the Queen of Sweden

In 1644 Descartes published the *Principles of Philosophy* enlarging his earlier work *The Meditations,* and presenting extra arguments dealing with their scientific application. His fame reached the ears of the 19-year-old Queen Christina of Sweden. He was to help her found a Swedish Academy and was to instruct her from 5 to 6 a.m. in philosophy.

The Queen, with her superhuman stamina, dragged him into her regimen of self-improvement, every winter morning. Descartes tried to collect his numbed wits and hold forth on the wonders of mathematics and philosophy.

But the inveterate 11 o'clock scholar could last just 11 weeks. He contracted influenza and died on February 11, 1650, in Stockholm. That year, his colleagues had already begun to fish out the bones of his work *La Geometrie* (*Time-Life Books,* 1984).

BLAISE PASCAL (1623-1662)

MATHEMATICIAN, THEOLOGIAN AND PHILOSOPHER

> He invented the first digital calculator, to help his father with the work of tax collection. This resembled a mechanical calculator developed in the 1940s.

Educated at home

Blaise Pascal was born at Clermont Ferrand (France) on June 19, 1623. His mother died when he was four. He was left with his two sisters, Gilbert and Jacqueline.

In 1631, the family moved to Paris where they went through a difficult period, having incurred the enmity of Richileu. He began attending meetings of a mathematical academy and had ***mastered Euclid's Elements of Geometry by the age of 12.*** Pascal's education was obtained within the family circle, his father teaching him Latin and Greek but not mathematics.

All mathematics books were therefore removed from their house. In 1641, the Pascal family shifted to Rouen where the elder Pascal was appointed tax collector. Pascal ***invented the first digital calculator*** to help his father with the work of tax collection. He sold in 1645 the first calculating machine. This resembled a mechanical calculator developed in the 1940s. There were problems faced by Pascal in the design due to the system of French currency at that time. Besides, ***his study of hydrostatics led to his invention of the syringe and hydraulic press (1647).***

Pascal had already written some of his best mathematical work by the time he shifted to Rouen. The intensity with which he worked was perhaps partly the cause of his ill health. As his health declined, ***his religious fervour increased.***

A religious experience

Despite the battle with his poor health, he continued to work assiduously until October 1654. Around his time, he nearly lost his life in an accident. The horses puling his carriage bolted and the carriage was left hanging over a bridge above the river Seine. Although he was rescued, he was affected psychologically. Not long after, he had a religious experience ***that altered the course of his life.***

He entered the Jansenist monastery at Port Royal, though he did not take orders. He began to write 18 epistles, under a pseudonym, a defence of his friend Antoine Arnauld, who was on a trial before the faculty of theology at Sorbonne (Paris) for heretical teaching. Known as *Provincial Letters,* dealing with religious topics, they were published from January 1656 to March 1651. In 1658, Pascal broke with the Jansenists and left the monastery.

Projective geometry

The mathematical theory of probability made its first great step forward in the correspondence between Pascal and Format (1601-1665), which consisted of five letters in the summer of 1654. Pascal planned a treatise on the subject, but only a fragment survived to be published after his death.

About half of Pascal's works are devoted to mathematics and physics. Before the age of 16, he had proved one of the important theorems in the geometry of conics, known as 'Pascal's mystic hexagram'. It is a theorem concerned with the collinearity of intersections of lines. Pascal used his theorem to derive 400 propositions bringing in the work of Apollonius and other geometers. Thus Pascal laid ***the foundation of Projective Geometry,*** an entirely underdeveloped branch of mathematics at that time.

Pascal then set to work on a book *Essays Oil Conics* and finished it in 1640. *The Generation of Conic Sections* was completed by March 1648; only a fragment of the work was published, the rest had been lost.

In his treatise on the 'Equilibrium of Liquids' (1653), he enunciated 'Pascal's Principle'. This is a complete outline of a system of hydrostatics, where he showed that ***in a fluid at rest the pressure is transmitted equally in all directions***. This was not at first accepted by his contemporaries; but now it is conceded that Pascal provided ***the essential link between the mechanics of fluids and the mechanics of rigid bodies.***

In October 1647, Pascal wrote about new experiments concerning vacuum, which led to disputes with a number of scientists like Descartes (1596-1650). The latter wrote rather cruelly to Huygens (1629-1695), after a two-day visit, "Pascal has too much vacuum in his head". But, Pascal showed successfully in August 1648 that ***the pressure of the atmosphere decreased with height and a vacuum existed above the atmosphere***.

The cycloid

Pascal's important contribution to mathematics after his entry to port Royal is said to have been the result of a toothache. To take his mind off the pain, he thought about ***problems associated with the cycloid,*** the curve traced by a point on the circumference of a rolling circle.

Though Pascal was not the first to write down the triangle of numbers, he was the first to use it in a theory of combinations and subsequently for calculating probabilities. Pascal worked intensively on the problem and solved it, evaluating the centre for gravity of the curve, the volume and surface area of the solid of revolution derivable from it. Pascal published his own solutions in the *Letters to Carcavi.*

However, he published a challenge to Wallis, Wren, Leibniz, Huygens and Fermat, offering two prizes for solutions to these problems. Wallis' solution was wrong; the others communicated their discoveries to Pascal without entering the competition.

Serving the poor

After this episode, he took little interest in science and spent his last years giving to the poor and going from church to church in Paris and attending one religious service after another.

His health deteriorated rapidly after 1658. He died at the age of 39, on

August 19,1662 in the home of his sister Jacqueline, who wrote a valuable biography of him.

Eight years after his death, an edition of his famous work in philosophy, *Pensees* (Thoughts), appeared. ***It is a collection of personal thoughts on human suffering and faith in God***. This contains ***the celebrated Pascal's wager:***

"If God does not exist, one will lose nothing by believing in him: while if he does exist, one will lose everything by not believing". (*Biographical Dictionary of Scientists,* Harper Collins Publishers, 1993)

GOTTFRIED LEIBNIZ (1646-1716)

WHO INFLUENCED MANY SCIENCES

He is the first mathematician to work out systematically the binary representation for any number. He felt a spiritual significance in this discovery: 1, representing unity, was clearly a symbol of God and 0, the nothingness.

Aptitude for self-study

Gottfried Leibniz was born on July 1, 1646 in Leipzig. He attended the Nicolai School, where his teachers struggled to confine the precocious boy to materials suitable to his age. A relative recognised the boy's aptitude for self-study and arranged for his free access to the library of his father Friedrick Leibniz, a professor of philosophy.

His universal genius influenced such fields as logic, metaphysics, mathematics, mechanics, geology, history, linguistics and theology. He dominated the intellectual life in Germany of the 17th century. For such spectacular achievements, he was largely ***self-taught.*** He had learned simple computation and a little geometry in his school. But his interest in mathematics was aroused by the numerous remarks on the importance of the subject that he encountered in reading philosophical works.

In 1661, he joined the University of Leipzig where he received most of his formal education in the Aristotelian tradition; he came into contact with the thoughts of men such as Galileo, Bacon, Descartes (1596-1650) who had revolutionised science and philosophy. Leibniz dreamed of reconciling these modern thinkers with the Aristotle School. In 1666, he wrote in *Acta* 'On the art of communication', and formulated in model, which is indeed the ***theoretical ancestory of the computer!*** — All reasoning, verbal or not, is reducible to an ordered combination of elements such as numbers, words.

Leibniz completed his legal studies in 1666, not yet 20; he was refused the doctorate degree and left his native city forever. But the University of Nurenberg made good by procuring him Doctor of Law as well as the immediate offer of a professor's chair.

Through J.C. von Boyenburg, a German statesman of the day, he met the Prince Elector, Johann Philipp von Schoenborn, who took him into his service. Leibniz worked on general legal problems. In 1672, the Elector sent Leibniz on a diplomatic mission to Paris. He met Christian Huygens (1629-1695), a Dutch mathematician, who first helped him with mathematics. By the time the mission was complete, Johann Philipp had died; his successor did not show interest to continue Leibniz's services. So Leibniz returned to Paris in March 1673, hoping with the reputation he had built, to get a salaried post in the Academy of Sciences.

Hanoverian period

As his efforts failed, Leibniz left for Hanover in October 1676, where he entered the service of the Duke of Braunschweig, John Frederick. He acted as adviser and librarian as well as consultant to various engineering projects in 1678; he was elevated to the post of Councillor.

Unlike Descartes (1596-1650), Leibniz had no fortune; again unlike Huygens (1629-1695), he did not get a pension. He had to ***earn a living by being a jack-of-all-trades to royalty.*** Leibniz worked on hydraulic presses, windmills, lamps, submarines and a variety of mechanical devices. He worked on a scheme, from 1680 to 1685, to increase the output of the Harz silver mines by using windmill-powered pumps. It failed, despite faultless engineering of the scheme. He also proposed a process for the desalination of water.

Work on a history of the earth

In 1685, Lebniz was named historian of the House of Brunswick. He did not restrict himself to a genealogy of the House but widened his goal to a history of the earth including geological events and description of fossils. He searched, in monuments and linguistics, for the origins and migrations of peoples; for the birth and progress of the sciences, ethics and politics; and finally for the elements of a *historia socra.* His research took him to various places in Europe; his princely support opened the door of archives and libraries; and he could meet eminent scientists.

Work in mathematics

These many diverse occupations did not stop his pursuit of mathematics. In March 1679, he perfected the binary system of numeration and also proposed the basis for general topology, a branch of mathematics that deals with selected properties of collections of related physical or abstract elements. In 1681, he was engaged in determining the proportion between a circle and circumscribed square, and later, he recognised the transcendence of the circular and logarithmic functions.

Leibniz published in the *Acta,* 'New method for the greatest and the least' which was an exposition of his differential calculus. It is known that Newton had discovered calculus as early as 1665, but he had not published his discovery. Leibniz's article opened up the question of ***the priority of the invention of calculus, a celebrated dispute of the era.***

Importance of inner relationships

In his approach to mathematical problems, Leibniz limited himself primarily to methods and results. He paid attention to formulating ideas clearly to connect them, as he did in an exemplary fashion in the mathematics of infinitesimals. Interesting details were important, but even more important were the inner relationships and their comprehension. He undertook the work needed by such an approach for purposes no other than the exploration of conditions under which new ideas are born, stimulate each other, and are joined in a unified thought structure.

Honours

On Leibniz's recommendation, the German Academy of Sciences was founded in Berlin in July 1700, whose president he became for life. He was elected to the Royal Society, London (1673). Lebniz's inventive powers and productivity in mathematics slowed down after 1700. Not all his manuscripts have yet been published.

Leibniz died in November 14, 1716, quite neglected by the Princes he had served. He died single.

JEAN ARGAND (1768-1822)

CELEBRITY BY A SINGLE INVENTION

His viewpoint to the geometric mean between two given vectors is an original result. It provided a new tool for research in geometry.

New precision to algebra

Jean Argand was born in Geneva, Switzerland, on July 18, 1768. He was a Parisian book-keeper, engaged in non-mathematical occupation. The details about his early education are scarce.

It is remarkable that Argand became a celebrity in the history of mathematics by his single contribution, ***the invention*** of a geometric representation of complex number which, in the words of the famous French mathematician Cauchy (1789-1857), ***gave "clarity, a new precision and a greater generality to algebra than earlier theories of imaginary quantities had".*** Apparently, Argand never belonged to any group of amateur mathematicians.

It is said that Argand's work was actually anticipated in 1797 by Casper Wessel, a Norwegian, whose work remained unknown until 1897. It might have met the same fate of obscurity, had it not been for the honesty and generosity of J.E. Francais, professor at the Ecole Institute of Artillerie who saw Argand's work mentioned in a letter amongst his brother's papers. The ideas in it were so intriguing that he developed them further and published them in Gergonne's journal *Annales De Mathematiques.* In this article, Francais expressed the hope that the unknown first author of these ideas would make himself known and publish his work.

Argand responded to the invitation by contributing a paper in the same volume of the *Annales.* A key to his ideas was presented in a figure giving a geometric representation of $\sqrt{-1}$; Argand followed John Wallis' (1616-1703) *Treatise of Algebra* (1685), since this quantity is the mean proportional between +1 and –1, its geometric representation could be a line constructed as the mean proportional between two oppositely directed unit segments. Argand employed the notation of bar over letters, to indicate what is today called a vector.

Work on negative numbers

In his book *Essai* on imaginary quantities, Argand cited several models for generating negative numbers by repeated subtraction. Initial use of the word 'imaginary' for a negative number may seem like nothing new, but ***it was a mathematical debate of that time whether negative numbers were numbers, or even existed.*** Argand used imaginary for multiples of $\sqrt{-1}$, a practice followed till today.

Argand ***coined the term Modulus*** for the absolute value or the length of the vector representing a complex number. Argand's notation is particularly significant because it anticipated the more abstract and modern ideas, later expounded by W.R. Hamilton, "of complex numbers as arbitrarily constructed new entities defined as ordered pairs of real numbers". This modern aspect of Argand's original work has not been generally recognised.

A new tool for research

Argand generalised the insertion of geometric means between two given vectors to the insertion of any number of means *n*, between the vectors by dividing the angle between them by *n*. By ***this purely geometric interpretation and reasoning, he arrived at the well-known theorem of De Moivre*** (1667-1754):

$$\text{Cos } na + i\, n \text{ Sin } na = (\text{Cos } a + i \text{ Sin } a)^n$$

where $i = \sqrt{-1}$

The above geometric viewpoint is an original result, and it ***provided a new tool for research in geometry.*** Argand was, no doubt, aware of the non-rigorous nature of his reasoning, but he defined his goals as clarifying thinking about imaginary quantities by setting up a new view of them.

Deriving infinite series

Argand used complex numbers to derive infinite series for several trigonometric and logarithmic functions. Argand's proof of what Gauss termed the fundamental theorem of algebra involved considering the modulus of

$$P(x) = a_0 x^n + a_1 x^{n-1} + \ldots\ldots + a_{n-1} x + a_n$$

where $x = a + ib$

Argand asserted the corollary that the theorem held, if the coefficients (a_i) were complex. Argand's last article appeared in Annales (1816) and dealt with a problem in combinations. He devised the notation (*m, n*) for the combinations of *m* things taken *n* at a time and the notation *Z (m, n)* for the number of such combinations.

In the first edition of his book *Lectures on Quaternions* (1853), Hamilton used lengthy footnotes to establish conclusively and to give priority to the quality of Argand's contributions. Argand died on August 13, 1822, in Paris.

Friedrich Bessel (1784-1846)

The complete astronomer

It is difficult to map the sky from the earth, as the dense atmosphere of the earth distorts the position of stars. Besides, the earth is a moving platform. Earlier astronomers had tried to obtain astronomical data in their own way. It was Friedrich Bessel who re-evaluated these corrections and evolved a logical, systematic scheme for applying them which has been followed universally ever since. He was thus the greatest astronomer of the 19th century and is probably the most complete astronomer of all time.

Apprentice in a firm

Bessel was born on July 22, 1784 in Minden, Germany. At school, apart from his interest in mathematics and physics, he did not show any sign of extraordinary talent until he was 15. In 1899, he began a seven-year apprenticeship with a famous mercantile firm in Bremen.

In order to qualify as a cargo officer, he studied books on ships and navigation. The problem of determining a ship's position at sea with a clock and sextant stimulated his interest in astronomy.

He was content with the operational procedures for navigation and so began to study astronomy and mathematics. The determination of the longtitude of Bremen and the observation of the eclipse of a star by the moon were among his first accurate astronomical measurements.

A turning point

In 1804, Bessel computed the orbit of Halley's comet from the observations made by Harriott in 1607. His results were published in a professional journal and the article had the standard of a doctoral thesis. It marked a turning point in Bessel's life: from then on ***he concentrated on celestial mechanics.***

Later, Olbers, the proprietor of his firm, claimed that his greatest service to astronomy was having encouraged Bessel to become a professional astronomer. On his recommendation, Bessel joined in 1806 as an assistant in a private observatory in Lilienthal. In 1807, Olbers encouraged him to do a reduction of Bradley's observations of the positions of 3222 stars, which were made from 1750 to 1762 at the Greenwich observatory. This work was one of his greatest achievements.

Meeting Gauss

Bessel was appointed by King Wilhelm III as director of the new observatory built at Koenigsberg. The greatest mathematician of his time, Gauss (1777-1855) had met Bessel in 1807 and had recognised his unusual ability. He ***proposed, without formalities, the title of doctor be conferred on him,*** which was a prerequisite for a professorship. Bessel took up his post in May 1810 and remained here for the rest of his life, leaving Koenigsberg only occasionally. He pursued his research and teaching without interruption, finding relaxation in daily walks and in hunting. The 21 volumes of his handbook in German *Beobachtungen der Koenigsberger sternwarte* stand testimony to his prodigious industry.

Practical astronomy

Bessel's work, *Astronomical Investigations* (1841-42), contained methods to calculate solar eclipses, which is still in vogue; the development of the heliometer (an instrument which measures the angular separation between celestial bodies) and a study of atmospheric refraction. ***Bessel is the founder of the German school of practical astronomy,*** which was later expanded to astrophysics by Bunsen's and Kirchhoffs discovery of spectral analysis.

Bessel's discovery of stellar parallax was a landmark in the history of astronomy; ***it heralded the beginning of all studies of the stellar system,*** which depend on the knowledge of distances.

Bessel function

He is renowned also in the field of engineering through a special class of mathematical functions *(known as Bessel)* that ***have become an indispensable tool in applied mechanics, vibrations of membranes, television and physics.***

Later years

"Bessel was small and delicate and, in his later years, he appeared prematurely aged because of his markedly pale and wrinkled face. ***This appearance altered, however, as soon as he began to talk; then the force of a strong mind was evidenced in brilliant, rapid speech*** and his otherwise rigid expression revealed mildness and friendliness." (*Dictionary of Scientific Biography,* Vol II)

Bessel married in 1812 and had two sons and three daughters. The early death of both the sons and frequent spells of illness marred his happiness. From 1840, his health deteriorated. After two years of great suffering, Bessel died in Koenigsberg on March 17, 1846. He was buried near the observatory.

HENRIK ABEL (1802-1829)

AN IMMORTAL IN GROUP THEORY

Abel came under the tutelage of Helmhoe, who changed his career. The two together started to study the texts of Euler, Lagrange and Laplace. Abel's progress was so rapid that he soon became the teacher!

Early life

Henrik Abel was born on August 8, 1802, in Finnoy, Norway in a family of seven children. His father, Georg Abel, was a Lutheran minister, gifted and a highly ambitious theologian. He was prominent in the nationalist movement for the creation of separate Norwegian institutions — particularly a university and a national bank, as a forerunner for outright independence.

Henrik Abel and his brothers were tutored at home till 1815 by their father, after which the former was sent to the Cathedral School in Oslo. Probably, the deteriorating family circumstances hastened his departure from home at the age of 13. During the first two years his performance was satisfactory; then it declined.

Guest in professors' houses

In 1817, ***he came under the tutelage of Holmhoe, who changed Abel's career.*** The teacher, discerning the boy's extraordinary ability in mathematics, began assigning special problems and recommending books outside the school curriculum. The two together started to study the texts of Euler (1707-1783), Lagragne (1736-1813) and Laplace (1749-1827). Abel's progress was so rapid that he soon became the teacher. ***His notebooks, which have been preserved in the university library, reveal*** that even at that time he was interested in the theory of algebraic equations.

Abel's father died in 1820, leaving the family in very straitened circumstances. He entered in 1821 the university and started his studies penniless, as there were then no funds for awarding fellowships. So some of the ***professors came forward to support the young Abel and to keep him as a guest in their houses.***

After a year, when he completed the requirements for the preliminary degree for doctorate, Abel devoted his time to advanced research on his own. He mastered enough of the mathmatical literature of the day. He published (1823) his first article on functional equations in a scientific journal started by Professor Hansteen, a leading scientist in the university. His second paper, which contained the first solution of an integral equation, went unnoticed. In the summer of 1823, he took a trip to Copenhagen to meet Degen and other Danish mathematicians, which gave him the much needed inspiration.

The quintic equation

On his return to Oslo, Abel continued his efforts for solving the quintic equation. He now hit on the reverse view and succeeded this time, by proving the impossibility of deriving a radical expression to represent a solution of the general fifth degree equation. Abel realised the importance of his result and had it published at his expense; and to reach a larger audience he wrote it in French. The whole matter was packed in six pages and the resulting brevity probably made it difficult to follow. Abel sent a copy to the great mathematician Gauss (1777-1855) but it did not elicit any response. Abel had ***resolved a 300-year-old problem,*** which is regarded now as a major result in algebra.

During the summer of 1825, Abel went on a study tour to learn foreign languages, for which he was awarded a grant for two years. He studied particularly French and prepared a considerable number of papers to be presented to foreign mathematicians. Hansteen advised him to spend his time in Paris, then the world's principal center of mathematics. But Abel decided to go to Berlin, as he could secure a letter of recommendation to Leopold Crelle, an influential Privy Councillor, who started the *Journal for Pure and Applied Mathematics*, which was the leading periodical in the nineteenth century.

The first volume alone of Crelle's journal contained seven papers by Abel, among them the expanded version of the proof of the impossibility of the solution of quintic equation by radicals. During that winter, Abel contributed notable papers — generalisation of the binomial formula and intergration of square root expressions.

Efforts in Paris

In July 1826, Abel reached Paris. His visit proved not rewarding, as the French mathematicians were aloof and not easily approachable. It was only for a short while that he met the legendary mathematician Legendre whose interests, in his old age, was elliptic intergrals, Abel's own speciality.

Abel completed during his stay in Paris, ***his great memoir,*** one of his longest papers, ***on the theory of algebraic functions*** which he presented to the French Academy on October 30, 1826. Cauchy (1789-1857) and Legendre (1752-1833) were appointed referees. No report was forthcoming, as Cauchy alleged that the manuscript was illegible. It was unfortunately lost. Abel's next two months in Paris were gloomy, as he had little money and few friends. He returned to Berlin, spending his scarce resources to pay the fare; shortly after, he suffered the first attack of tuberculosis. Crelle tried to keep him in Berlin, until he could find a position for him in a German University; he graciously offered him the editorship of his journal.

Theory of elliptic functions

Abel worked assiduously on a new paper and extensive publication on the theory of elliptic functions, which became a natural generalisation of the trigonometric functions. He was able to present the theory with rich details, including double periodicity, expansions in infinite series and addition theorems, as he had already worked out most of the theory as an undergraduate student. Abel's original work pushed the theory of elliptic functions to the frontiers of mathematical research during the nineteenth century.

Abel returned to Oslo on May 20, 1827, only to find that the conditions were gloomy, no academic prospects, no fellowships but mounting debts. He earned his livelihood by giving tuition to school boys and with some support from his friends.

In spite of his deteriorating health, he wrote new papers frantically. The first part of 'Researches' was published in Crelle's journal in Septmeber 1827; the second part was completed by Abel in the winter. He did not know that a competitior, Jacobi (1804-51) by name, had appeared in the field of elliptic functions until early in 1828. Jacobi acknowledged that in order to derive some of his results he had to rely on Abel's paper. On June 28, 1830, ***the French Academy awarded its Grand Prix to both Abel and Jacobi*** for their outstanding mathematical discoveries.

The great memoir already referred to, ***was rediscovered in Paris and published in 1841. During printing it again disappeared; it reappeared in 1952 in Florence!***

Abel remained in obscurity and suffered financial troubles. He was unable to secure a regular university post and died of tuberculosis in the prime of his youth on April 6, 1829: two days after, a letter arrived appointing him professor at the University of Berlin! His name is ***immortalised in Group Theory.***

Tribute

Crelle wrote a moving tribute in his *Journal of Pure and Applied mathematics:* "All of Abel's works carry the imprint of an originality and force or thought which is unusual... It was not only his great talent which made his loss infinitely regrettable but he distinguished himself equally by the purity and nobility of character and by rare modesty, which made his person to the same unusual degree as was his genius".

(*Dictionary of Scientific Biography,* Vol. I, Ch Scribner's Sons, New York.)

The Abel Prize

It was founded in 2001 to commemorate Abel's bicentenary, for original achievements in the fields such as statistics, number theory and the mysteries of prime numbers which find application in computer science. It is considered equivalent to the Nobel Prize.

The first prize was awarded at Oslo in 2003. Prof. S. R. Srinivasa Vardan from India (Director, Courant Institute of Mathematics, New York), received the award in 2007. The Norwegian mathematical wizard is sometimes compared to our mathematical genius Srinivasa Ramanujan.

EVARISTE GALOIS (1811-1832)

WHO HERALDED THE BIRTH OF MODERN ALGEBRA

There have been few mathematicians who can equal the contributions of astonishing richness that Galois could accomplish in a short span of life of 20 years, spent in frustrating circumstances and driven to revolt. Continuing the work of Abel (1802-29), Galois produced, with the aid of group theory, a definitive answer to the problem of solvability of algebraic equations. In this, he was undoubtedly the beneficiary of the works of his illustrious predecessors Cauchy, Gauss and Lagrange, but his brilliant sense of mathematical thinking made him an ***exceptional innovator.***

Seeking refuge in mathematics

Galois was born on October 25, 1811, in a village near Paris. He was taught by his mother until the age of 12; then he entered a boarding school in Paris. His first contact with mathematics was in 1827, through a preparatory course given by his teacher Vernier. ***The rigour of mathematics was a revelation for him.***

His untameable genius for original work showed up in his school years, as he turned to reading the orignal works of the celebrated mathematicians Legendre (1752-1833) and Lagrange (1736-1813), which gave him a solid grounding. Following his father's suicide, ***he sought refuge in mathematics,*** studying contemporary literature on the theory of equations and elliptic functions. The results he obtained in May 1829 were communicated to the French Academy of Scienes by Cauchy (1789-1857). This paper that he wrote at the age of 18 was lost.

A month later, he failed in the entrance examination for Ecole Polytechnique. He was obliged to settle for the second best, joining in November 1829 the Ecole Normale Superieure to be trained as a teacher. Through reading *Ferussac's Bulletin of Mathematics,* he learned that Abel died just then and that a good number of results he had presented in his paper to the Academy were already contained in Abel's last memoir.

Work in the face of personal risk

Galois then worked on a revision of this paper, which he submitted to the academy in February 1830, hoping to win the Grand Prix in mathematics. Unfortunately, this memoir was also lost, on the death of Fourier (1767-1830), who was to be the examiner. However, a brief report was published in the April 1830 issue, of the *Ferussac's Bulletin,* which carried the original works by Cauchy (1789-1857) and Poisson (1781-1840). This itself is sufficient testimony to the reputation that Galois had then acquired. By November 1830, Galois had taken up the Republican cause. He published a violent article criticising the director's action, for which he was expelled from Ecole Normale in December 1830.

Tragic duel

In the political turmoil following the French Revolution, he was arrested on July 14, 1831, and sent to prison where, in spite of a troubled and painful situation, he revised his memoir on equations and worked on the application of his theory. He was released on parole on March 16, 1832, but was soon challenged to a duel presumably caused by a love affair, by an unknown adversary.

On May 29, 1832 he spent the night hastily sorting his papers and sketching out, in a testamentary letter intended for Jacobi as many of his mathematical discoveries as he could, occasionally breaking off to ***scribble in the margin "I have not time".*** At, dawn he received a pistol shot and, having been left where he fell, was found dead by a passing peasant (*Chambers Concise Dictionary of Scientists,* W&R Chambers Ltd., Edinburgh, 1989). He died under tragic circumstances on May 31, 1832. However, the testamentary letter referred to already, was published in September 1832. It did not appear to have attracted the attention it deserved.

Discovery of group theory

Galois had built on an idea by Lagrange in 1770 on permutation of the solutions of equations; using the idea and some of Abel's results, Galois discovered ***a remarkable theory linking the roots of an equation to the properties of the permutation of the roots.*** If the roots obey the same relations after permutation, they from a Galois group and this gives information as to the solvability of the equations. The few fragments published during Galois' short life were hardly noticed by his contemporaries. In fact, a paper on the general solution of equations was sent to the Academy of Sciences through Poisson but was rejected as incomprehensible.

It was Liouville (1809-82) who announced in September 1841 officially to the Academy that Galois had effectively solved the quintic problem. He published Galois manuscripts in *Journal of Pure and Applied Mathematics,* October 1846.

Influence on modern algebra

The originality of Galois' conceptions and the extreme conciseness of his arguments made his works accessible only to the best among the mathematicians of his time. His contemporaries treated his works with disdain and contempt. It was not until the publication in 1870 of Camille Jordan's book that group theory and the whole body of Galois' writings became available to mathematicians. From that time on, ***Galois' work exerted a profound influence on the development of modern algebra.***

Galois expressed in succint words the task ahead: "Jump with both feet of the calculus and group the operations, ***classifying them according to their difficulties and not according to their forms;*** such, in my view, is ***the task of future mathematicians".***

George Boole (1815-1864)

Founder of mathematical logic

The design of the modern digital computer owes much to a concept developed some 110 yeas ago by a self-taught mathematician, George Boole. ***The transition from algebra as letters-for-numbers to algebra as structure was a revolution*** in mathematics in the last century. Boolean Algebra has been applied to the design of logic circuits with simple switches and control elements, the basis for computing systems.

Following the footsteps of Aristotle and Leibniz, Boole founded mathematical logic to show how a symbolic notation, analogous to that used in algebra could be employed to express and generalise logical processes.

Love for mathematics

Boole was born in Lincoln, Ireland on November 2, 1815. He apparently inherited his love for mathematics from his father, a cobbler whose interests lay in mathematics and making optical instruments. Due to a setback in his father's business, ***he had to stop school education early***.

Boole then resolved to become a clergyman: the four years for this career were not wholly wasted. He had acquired a mastery of French, German and Italian, all destined to be of indispensable service to him.

His career

At the age of 15, he took to teaching and soon set up a school in Lincoln. Events took a turn for the better in 1834, when ***the Mechanic's Institute was founded in Lincoln,*** giving everyone access to books. Boole became the curator of the institution's reading room, which provided him with publications of the Royal Society. Though running of his school and the curator's job took most of his time, he devoted his scanty leisure to the study of mathematics. As a schoolmaster, he was sought out by eminent mathematicians such as Arthur Cayley (1821-95).

Boole grappled with Newton's, Laplace's and Lagrange's mathematical treatises. When he was only 19, he was invited to give a commemoration address at the presentation of a bust of Newton to the Institute. He published in 1847 *The Mathematical Analysis of Logic* wherein ***he asserted that logic did not form part of philosophy but was associated with mathematics.*** He also applied his ideas to the theory of probability.

His scientific writings elevated him, ***with no university degree, to the post of Professor of Mathematics*** at Queen's College, Cork, in 1849. Despite the teaching load, he found time for research and wrote his celebrated work *An Investigation of the Laws of Thought* (1854), mature statement of his ideas. ***It is a classic in the history of mathematics, establishing both formal logic and the algebra of logic.*** "Pure mathematics was discovered by George Boole", said Bertrand Russell.

In the first book, he objected to the current view of mathematics as the science of magnitude or number, amplified and clarified the ideas he presented. He invented a radically different form of algebra based on the algebra of sets. This culminated in the second book (1854).

What is Boolean Algebra?

Boole showed that his algebra provided an easy algorithm for syllogistic reasoning and that the five fundamental laws of algebra now hold for this algebra:

$$x + y = y + x, \; xy = yx$$

$$x + (y + z) = (x + y) + z, \; x(yz) = (xy)z$$

$$x(y + z) = xy + xz$$

Nevertheless, not all the rules of ordinary algebra continue to be valid. For example,

$$1 + 1 = 1 \text{ and } x.x = x$$

Boole's work prompted a sequence of axiomatic studies by C. S. Pierce (1839-1914), who greatly enlarged Boolean algebra by the introduction of a new kind of abstraction.

A committed teacher

His two books on 'Differential Equations' and 'Calculus of Finite Differences' were standard textbooks for many years in the U.K. He was instrumental in ***introducing the continental approach to calculus, and teaching methods in Britain.*** In 1857, he was elected Fellow of the Royal Society (FRS).

Boole was a conscientious and committed teacher. Such was his commitment that he walked in rain and lectured in wet clothes, which led to illness that resulted in his death on December 8, 1864.

His wife, Mary, shared his interests in mathematics, as she developed some remarkable insights into methods of teaching and learning.

(*A History of Mathematics,* Boyer and Merzbach, John Wiley and Sons, New York, 1991)

YASUDAS RAMACHANDRA (1821-1880)

MATHEMATICIAN AND EDUCATIONIST

The introduction of modern science in our country a century ago triggered a number of cognitive encounters which were often viewed as mere conflicts between different knowledge systems, in particular Eastern and Western. One of the forms which this encounter shapes, was the revitalisation of traditional knowledge systems by the avant-garde of the Indian intelligentsia.
Here is one such cognitive encounter, the efforts of the little-known polymath and Urdu translator Master Ramachandra ***who saw in Algebra the greatness of Hindu intervention.*** He developed a new method to solve all problems of maxima and minima by algebra and sought to introduce the Indian people to the latest developments in calculus in their local language.

Self-taught

Ramachandra was born in 1821 at Panipat, near Delhi. His father, an official of the East India Company, died in 1831 when Ramachandra was just nine years. Therefore, he was brought up and educated by his mother. He was admitted in 1833 to an 'English school' in Delhi, where he remained for six years. ***He had to teach himself mathematics, as the English school had no facility to teach that subject.*** He studied at home with books that he bought with his scholarship: this early preparation brought him closer to the indigenous perception of mathematics and groomed his rather unconventional knowledge in the subject.

Ramachandra was married at the age of 11 to a girl who, he later found, was deaf and dumb. Despite financial difficulties and the difficulty of caring for an invalid wife, he pursued mathematics with devotion.

His career

In 1843, Ramachandra taught science through the medium of the vernacular, at Delhi College. This led to his being called 'Master Ramachandra'. Besides, he was entrusted by the principal, Mr. Boutros, with the project for translating English scientific works into Urdu. These activities were formalised under the aegis of the Vernacular Translation Society. Articles on new inventions, discoveries in modern science and technology followed at the behest of the next principal Dr. Sprenger, in the two Urdu papers *Fawaid-ul-Nazrin* and *Qiran-us-Sadain.* He became a prolific writer in Urdu, which career he pursued assiduously. The achievements of Western civilisation and the scientific temper of Europe inspired Ramachandra.

Teaching of science in vernacular

Ramachandra advocated that the local language should be the only medium of instruction. Sir Syed Ahmad (Aligarh), his contemporary, supported this contention.

So Ramachandra took up the ***translation of a number of important English books on science into Urdu***, to enable teaching of the subject in the local language. Some of the titles are: *Plane Trigonometry* (1859), *An Elementary Treatise on the Theory of Equations* (1861), *The Elements of Euclid* (1862), *Algebra for Beginners* (1863), *Galbraith and Haughton's Scientific Manuals,* 4 Vols (1859-65).

Popular articles on science

In his capacity as an Urdu journalist, Ramachandra wrote regularly on subjects of scientific interest, namely on the 19th century developments in science and technology. In his journal *Fawaid-u1-Nazrin,* where he imitated the plan of the legendary *Spectator,* he wrote popular articles on science. Besides, he published a series of articles on men of science like Elucid, Galileo, Lagrange, Laplace, Newton and Pythagoras.

Science was being introduced in schools by evangelicals like Grant and

Wilberforce to undermine religions and customs. Indian intellectuals found in science a neutral pursuit, which could serve for articulating a counter-colonial political stance. The role science came to play was for bringing about societal change and for opposing the cultural and intellectual erosion unleashed under colonial rule. For Ramachandra and the generation of Indian scientists that followed, ***science served not only as a catalyst of change but also as a weapon in the struggle against imperialism.***

The two books

Ramachandra wrote his famous book, *A Treatise on the Problems of Maxima and Minima* in 1850, when he was just 29 years old. It was severely criticised in the *Calcutta Review,* to which he sent a rejoinder. It ***was published again in 1859 from London*** at the initiative of the English mathematician De Morgan.

The cover of the British edition (1859) bears the following notification:

"Reprinted by order of the Honourable Court of Directors of the East India Company, for Circulation in Europe and in India, in acknowledgement of the merit of the author, and in testimony of the sense entertained of the importance of independent speculation as an instrument of national progress in India.

Under the Superintendence of Augustus De Morgan of Trinity College, Cambridge."

The first copy was presented to Lord Lyndhrust by the Secretary of State for India. ***De Morgan wrote a scholarly introduction*** to the book, where he recounted on what then was existing in our country:

Glory of Sanskrit

"There still exists a body of literature and science which might well be the nucleus of a new civilisation though every trace of Christian and Mohamedan civilisations was blotted out of existence. ***There exists in India, a philosophical language Sanskrit, which is one of the wonders of the world,*** and is a near collateral of Greek, if not its parent form. From those who wrote in this language we derive our system of arithmetic, and the algebra which is the most, powerful instrument of modern analysis: an astronomy worthy of comparison with that of Greece in its best days: above comparison, if some books of Ptolemy's Syntaxis be removed."

Ramachandra wanted to solve all problems of maxima and minima with algebra "to resusticate the native desposition of Indians that had been eroded over the centuries". He asserted that his was a new method, namely making an algebraic function, a maximum or minimum, without introducing the concept of differentiation. This is testified in De Morgan's introduction:

Ramachandra derived ***his method starting from Bhaskara's 'Bija Ganita',*** his ambition was to develop pedagogy for teaching mathematics in a familiar tradition.

His second book, *A Specimen of a New Method of the Differential Calculus* (called the Method of Constant Ratios), was published in Calcutta in 1863. The methods developed in both books were seen as pedagogic variations.

Later career

Ramachandra embraced Christianity in 1852. His life was in danger during the 1857 mutiny. He had to flee for safety and stayed in a village with a zamindar.

From September 1857, he served as headmaster at the famous Thomason College of Civil Engineering, Roorkee, (U.P.), the first of its kind founded by the British in our country. In September 1858, he shifted to Delhi as the headmaster of the Delhi District School. He retired because of ill health, in 1866. Soon after, he became tutor to the Maharajah of Patiala and director of education in 1870. The Maharaja honoured him with a jagir for his contribution to the development of education in the state. Ramachandra's health deteriorated with age and he died on August 11. 1880 at the age of 59.

Ramachandra's two mathematical treatises, his popular articles on science carried as a regular column in his Urdu journals, his translation of scientific works from English to Urdu and ***his efforts in promoting the teaching of science in mother tongue, have received scant attention from historians of science.***

(National Institute of Science, Technology & Development Studies, New Delhi 110 012)

Leonhard Euler (1707-1783)

Against all odds

For mathematicians, the 18th century is generally called the Age of Euler. But his influence went far beyond, as the work of many 19th century mathematicians had their roots in the works of Euler. For instance, one can trace back to Euler numerous paths from the mathematical school of Chebyshev of St. Petersburg University.

Leonhard Euler was, in fact, the creator of concepts and techniques. He introduced ***many current notations*** such as the sigma for the sum; the letter *f* and parenthesis for a function; *e* for the base of natural logarithms; $\Delta y, \Delta^2 y \ldots$ for finite differences; modern signs for trigonometric functions and *i* for square root of -1. His numerous discoveries — the theory of ordinary and partial differential equations, theory of multiple integrals, the rule of substitution, calculus of finite differences — enriched mathematics to solve problems in technology.

Early education

Euler was born on April 15, 1707 in Basel, Switzerland. His father, who was deeply interested in mathematics gave him elementary education at home. Euler studied in a local gymnasium where mathematics was not taught. So, he privately studied difficult mathematical books. He was introduced to the famous professor, John Bernoulli (1667-1748), about whom he says, in his brief autobiography (*Dictionary of Scientific Biography,* Vol. IV. Ch. Scribiner's Sons, New York), "He refused flatly to give me private lessons; but gave me permission to visit him every Saturday afternoon and he kindly ***explained to me everything I could not understand****....* and this, undoubtedly, is ***the best method to succeed in mathematical subjects".***

At St. Petersburg Academy

In the summer of 1722, Euler delivered a speech on 'temperance', for which he received a degree equivalent to the B.A. He competed in 1727 for a vacancy at the University of Basel; but, as posts were filled by drawing lots, he was not among the lucky successful candidates. This, however, turned out to be his good fortune, as he received an invitation to serve an associate of physiology from St. Petersburg Academy of Sciences. He became professor of physics in 1731, succeeding Daniel Bernoulli (1700-82). He found himself in the company of eminent scientists, which played a role in the flowering of his genius.

Euler had several tasks to carry out for the government and to find solutions for various technological problems: studying Russian territory and working on maps, studying problems of ship building and navigation.

During 1727-1741, Euler made brilliant discoveries in areas such as analysis, theory of numbers and mechanics. And by the age of 34, he had 55 publications including the two-volume treatise *Mechanics.* His quiet and productive career received a severe shock when he lost his right eye in 1738. This was compounded by the degeneration that set in the St. Petersburg Academy. So, accepting an invitation, he shifted to the Berlin Academy, in 1741.

At Berlin Academy

Euler worked in Berlin for the next 25 years. He carried through a formidable programme of work: supervision of the observatory; botanical gardens; the organisation of lotteries; publication of various calendars and maps (which earned income for the Academy); studies on demography and insurance. But these many occupations did not stop him from pursing the main vocation, ***the teaching of mathematical sciences and preparation of university texts***.

During 1765 and 1766, grave conflicts arose with King Frederick about the management of the Academy after the Seven-Year War and Euler left

Germany. Being unusually prolific, he had prepared no fewer than 380 works, including seven books, during his stay in Berlin.

Calamities he faced

Euler returned to Russia in July 1766. Soon after his arrival at St. Petersburg, he suffered from a cataract in his left eye. By 1771, ***he became totally blind.*** He suffered another calamity: the loss of his house and his personal property in a fire, barely managing to rescue himself and his manuscripts.

Undaunted, he devised new plans to continue his work. He involved his two sons, Albrecht and Christoph, and two academicians, Kraft and Lexell, as active collaborators. Being blind, he dictated his works. But ***his productivity continued undiminished,*** sustained by a flawless memory and a remarkable capacity for mental computations.

Creating mathematical science

Euler had only a few disciples, none of them was a first-rate scientist. They, however, formed the first mathematical school and contributed to the development of mathematical education in Russia.

By his life and work he, after Newton, created mathematical science and elaborated methods for the direct solution of problems in fields as varied as mechanics and astronomy, physics and navigation, geography and geodesy, hydrodynamics and fluid mechanics, insurance and demography.

On September 18, 1783, Euler died in St. Petersburg of brain hemorrhage.

JOSEPH LAGRANGE (1736-1813)

THE SELF-TAUGHT GENIUS

Napoleon elevated Lagrange to the rank of a Count. But, he remained the quiet and unobtrusive academician, living only for science.

Turning to mathematical analysis

Joseph Lagrange was born in Turin, Italy, on January 25, 1736. At the College of Turin, he studied classics and planned to study law, a profession that one of his brothers had pursued. But a ***chance reading of a memoir of the English astronomer Edmond Halley*** (1656-1743) aroused Lagrange's interest in mathematics. The memoir dealt with the use of analytical methods in the solution of problems in optics. He soon became aware of his talents and turned to mathematical analysis, then a rapidly developing field.

His father was treasurer to the King of Sardinia and lost his fortune in financial speculation. Lagrange is believed to have remarked, ***"If I had been rich, I probably would not have devoted myself to mathematics!"***

In 1755, he was appointed, when he was only 19, professor of mathematics at the artillery school of Turin. Two years later, he helped to form a society, which became the Turin Academy of Sciences. His papers published during this period, on the propagation of sound and on the concept of maxima and minima, won him praise from the famous mathematician Leonhard Euler

(1707-1783) in Berlin. Thus by 1761, he was already recognised as one of the greatest mathematicians.

Work at the Berlin Academy of Sciences

In 1764, he was awarded a prize by the Paris Academy of Sciences for an essay on the liberation of the moon. He won the same distinction in 1766 for the theory of the motions of the satellites of Jupiter and again in 1772 for the three-body problem. In 1776, Lagrange moved to Berlin to take up the post vacated by Euler at the Berlin Academy of Sciences. This was at the invitation of Frederick the Great, who expressed "the wish of the greatest King in Europe to have the greatest mathematician in Europe at his court".

Lagrange worked in Berlin till 1787, publishing papers on a wide range of subjects — the three-body problem, prime number theory, differential equations, calculus of probabilities, celestial mechanics, and the solar system. He opened a new period in algebra, which inspired Galois to formulate his 'group theory'.

Shifting to Paris

With Frederick's death in 1786, Lagrange opted to accept the invitation of Louis XVI. He stayed in Paris till his death in April 10, 1813. This period was dedicated to didactic writings and to the composition of his treatises summarising his mathematical conceptions: these treatises, while closing the age of 18th century mathematics, prepared and in certain respects, opened that of the 19th century.

Lagrange had left examples to follow, new problems to solve and techniques to develop in all branches of mathematics. His works were published in 14 volumes.

Calculus of variations

i. From the Louvre, where he was given apartments, Lagrange published his classic *Mecanique Analytique* (1788), a lucid synthesis of a century of research in mechanics since Newton based on his own calculus of variations. This led to the formulation of independent coordinates necessary for specifying a system of finite number of particles, now called the 'generalised coordinates'. The book was typically analytic: ***he remarked in the preface that no diagram would be found in the book but only algebraic equations***. This became the textbook on which all later work in the field of analytical and celestial mechanics would be based.

ii. Lagrange's lectures were published in two volumes, in 1797 and 1804, the first textbooks on real analytical functions. Besides, he developed the theory notation and application of the calculus of variations in a series of memoirs. The basic ideas were not fully grasped by his contemporaries, as no rigorous justifications of the principles were given. But the results amply vindicated the method.

Career in Ecole Polytechnique

The Revolution that began in 1789 pressed Lagrange into committee work, to reform the metric system and then into teaching. So when the Ecole Polytechnique was opened in Paris in 1795, he was appointed professor along with his friend, G. Monge (1746-1818): ***the two mathematicians who complemented each other very well became the most famous teachers in Paris*** for the next generation of French mathematicians.

The mathematician who most influenced Lagrange was Euler (1707-83), whom he had never met. Yet, even in the face of this great model, he preserved an originality that allowed him to criticise, to systematise and to deepen the ideas of the older mathematicians.

Napoleon honoured Lagrange, making him a Senator and elevating him to the rank of a Count. But Lagrange remained the quiet and unobtrusive academician, ***living only for science.***

Pierre Laplace (1749-1827)

Who transformed mathematical analysis

"It is India that gave us the ingenious method of expressing all numbers by means of ten symbols, each symbol receiving a value of position as well as an absolute value: an important and profound idea which appears so simple that we ignore its true merit. But it's very implicit and the great ease which it has lent to all computations put out arithmetic in the first rank of useful inventions."
— Laplace

How he entered the mathematical world

Laplace came from a poor family: he was born on March 23, 1749 at Beaumont in Normandy. His father, a peasant farmer, intended him for an ecclesiastical vocation. The young Laplace, matriculated in 1766 from the University of Caeny and worked for some time as a teacher. Determined to make his way in mathematics in 1768, he left his humble surroundings for ever and departed for Paris to conquer the mathematical world. He carried with him a letter of recommendation to D'Alembert (1717- 83), the famous mathematician. The latter gave him a problem and a week's time to solve it. Legend has it that Laplace solved it overnight and D'Alembert secured for his new protege the post of professor of mathematics at the 'Ecole Militaire'.

In 1773, Laplace won the election to the Academy of Sciences and, in just three years, published 13 papers on a wide range of topics. He was appointed in 1784 as examiner of cadets in the military school. ***He met here the young Napoleon whom he examined in mathematics.*** This youth was destined to deflect him from his avowed devotion to mathematics into the muddy waters of politics.

Laplace's scientific career can be considered to have taken shape in four stages, his contributions being pretty much in that order.

Advance in physical astronomy

i. 1768-78: he wrote memoirs on problems of integral calculus, astronomy, cosmology, games of chance and causality. During this formative period, he established his style of working and a programme of research in two areas, celestial mechanics and probability.

In 1773, he threw himself into his life-work — ***the application of the Newton's law of gravitation to the entire solar system*** by accounting for all of the observed deviations of the planets from their theoretical orbits and developed a conceptual view of evolutionary change in the physical universe. This discovery, which led to establishing the stability of the solar system, was the most important advancement in physical astronomy since Newton.

ii. 1778-89: In this period, he developed certain mathematical techniques, notably generating functions, the transform since known by his name and the theory of potential.

From Lagrange (1736-1813), it is said, he lifted the fundamental concept of the potential. ***Out of this theory grew the mathematics of boundary-value problems; it made possible an attack on physical problems*** which otherwise would have been unsolvable.

Publication of the three treatises

iii. 1789-1805: This period propelled him to the zenith of his career. He saw the publication of his three great treatises 'Analytique de probabilities', 'Mecanique celeste' and 'System du monde'. Laplace revealed himself almost as great a writer as a mathematician.

The first appeared in four volumes from 1799 to 1805, and dealt with the generalisation of the laws of mechanics for their application to the motions and figures of the heavenly bodies. In the second treatise, Laplace's contribution may be characterised as the extensive application of analysis in probability theory. He developed a method of approximation for definite integrals with integrands containing factors of high powers, a type frequently appearing in probability theory. Laplace was also the first to apply infinitesimal calculus to problems in probability. The third appeared in 1796 (two volumes). The topics treated were perturbations in planetary motions, the shape of the earth, motions of the tides, lunar motions and the rings of Saturn. Laplace developed mathematical methods, in a far more accurate manner than Newton, for calculating the disturbances of planets and moons and the forces which act according to the inverse square law.

He followed the text of his treatises by informal talks to intelligent audiences and worked on revising and updating them throughout his life. So, the *'System du monde'* won acclaim as 'one of the most successful popularisation of science ever composed'.

Research in physics

iv. 1805-27: The centre of his interest now shifted to physics — capillary action, namely the rise of liquids in narrow tubes, the formation of drops, heat, the theory of vapour and optics. He made the first advance toward the understanding of molecular forces in liquids and explained the inverse proportionality between capillary rise and diameter of tubes.

Laplace made a ***fundamental contribution in the theory of sound:*** the speed of propagation is influenced by heat. Laplace calculated this influence and found that the square root of the ratio of two specific heats of air (at constant pressure and constant volume) is to be added to Newton's formula.

It was during this period that ***he formed the famous society of Arucell and displayed his ability to dictate problems and solutions*** in the intimate atmosphere of the disciples he gathered. Thus the Laplacian school encouraged mathematisation of science and exerted a dominating influence on French physics. It was left for Poisson (1781-1940), his uncritical admirer, to carry on Laplacian physics into the late 1820s and 1830s.

Contribution to scientific education

Laplace made lasting contributions to education. His activities in the newly founded Institute de France raised the exact sciences to a paramount position. He introduced the system of prize competetions, notable among them being the one for a mathematical study of double refraction.

His involvement in the Ecole Polytechnique gave a new direction for scientific education at large. Laplace introduced the practice of conducting competitive examinations for selecting elitist rather than privileged students to enter this institution. From among them rose the stalwarts of the early nineteenth century — Ampere, Carnot, Fresnel, Poisson and Malus. Laplace served as examiner from 1795-99 and exercised authority over the content as well as quality.

Laplace was considered an opportunist, as his political allegiance shifted easily as expediency required. The successive revolutionary governments — the republic, the restoration and the empire, rewarded his servility with numerous honours.

Honours

He was elected in 1816 to the Academy of France. Napoleon made him a minister, which lasted only six weeks, as the former considered his record as an administrator undistinguished. Louis XVII made him a Marquis.

Pierre Laplace is among the most influential scientists of the eighteenth century. He is known for his technical contributions to exact sciences, for the philosophical point of view he imparted to the modern scientific disciplines and for his dominant role in shaping scientific education in France.

But throughout the troubled political times, ***Laplace's devotion to science remained unchanged.*** He spent his retired life comfortably at his country estate at Aruceil near Paris. He died on March 5, 1827.

JOSEPH FOURIER (1768-1830)

WHO SIMPLIFIED SIGNAL PROCESSING

"There was at the Academy of Sciences, a celebrated Fourier whom posterity has forgotten", wrote Victor Hugo, seeking a contrast with the growing fame of another Fourier, the socialist Charles Fourier. Joseph Fourier's work was not appreciated during his lifetime. He drew recognition in inverse proportion to history's estimate of their importance: as an able administrator, his studies in Egyptology, and ***his original contributions to mathematical physics.***

Scramble for life

Joseph Fourier was born in March 21, 1768 in Auxere (France) to a tailor. When he was only nine, his father's death left him an orphan; only by his intelligence could he gain a free place at the local military school conducted by Beneditine monks. He began to prepare for their teaching order but his plan was altered by the outbreak of the French Revolution. The Napoleonic era forced Fourier more than once to scramble for his life and his livelihood.

When the Ecole Polytechnique, a new college, was opened in 1795, he was appointed assistant lecturer to support the teaching of Lagrange (1736-1813) and Monge (1746-1818). His talent was soon recognised and in 1798, he succeeded Lagrange as professor of analysis and mechanics. This quiet interlude was interrupted by his service in Egypt from 1798 to 1801, as a member of the French expedition led by Napoleon.

A skilled administrator

On his return in 1801 to France, Fourier was offered by Napoleon the position of Prefect of Isere (equivalent to the governor of a province). This necessitated his exile from the academic circles at Paris to the remote Alpine city of Grenoble.

Two major achievements during his 14 years in office showed his skill as an administrator. The first was the draining of 20,000 acres of swamps between Lyon and Grenoble against the suspicions of peasants. He won them over by visiting them in their homes and impressing on them the advantages they would enjoy in the form of increased acreage. The second was the construction of a new road across the Alps connecting Grenoble to Turin in Italy. He circumvented the opposition by writing a brilliantly lucid paper on the advantages of his scheme, which won Napoleon's approval.

In 1809, Fourier started compiling and publishing the influential *'Description of Egypt'*, a 21-volume work which he completed in 1825. ***The work marked Egyptology as a new and separate discipline.*** Napoleon rewarded Fourier in 1809 by making him a Baron.

The low point in his career

In 1815, with the fall of Napoleon, Fourier gave up his post as Governor and returned to Paris to take up full-time research at the Ecole Polytechnique. This was the low point of Fourier's life: he had ***no job and low political reputation.*** But a former classmate and companion in Egypt secured him appointment as director of the Bureau of Statistics in Paris, a post without onerous responsibilities. This enabled him to devote his time to research and writings. Now, he was able to take his proper place in the centre of French intellectual society as the grand old man encouraging talent (Louisville, Sturm, Dirichlet). Finally he attained the summit of French distinction, member of the 'Academie Francaise'.

Impact on mathematical physics

Fourier strongly influenced mathematical physics through his treatise, *The Analytical Theory of Heat* (1822). He showed how the conduction of heat in solid bodies may be analysed in terms of infinite mathematical series, now known as the Fourier series. The obsession of Fourier for warmth, which Egypt satisfied possibly, explains ***his particular choice of propagation of heat as his research.***

His work had a great impact on the theory of functions of a real variable. Fourier insisted on substantiating his theoretical work by physical measurements in the laboratory, which should be of interest to application-oriented engineers of today.

Originality of Fourier's methods

In his approach to solve the partial differential equations, Fourier developed ***a set of mathematical techniques — 'separation of variables',*** where none had existed before. One wonders where the field of automatic control would now be without Fourier's discoveries.

"Unable to accept *Fourier's assertions concerning expansions in trigonometric series*, Laplace (1749-1827), Lagrange (1736-1813) and Poisson (1781-1840) attacked his theory of heat conduction as lacking in rigour. His contemporaries could not see into the future and their doubts are surely more a tribute to the originality of Fourier's methods than a reproach to the colleagues whom Fourier greatly respected." (*Fourier Analysis,* Koerner, Cambridge University Press, 1988).

During the last five years of his life, when recognition by his peers came to him, his health declined. Despite being afflicted with rheumatism and malaria, he did not stop his mathematical work. Some of his work during 1820s concerns problems in what is now called 'linear programming'.

A jungle without pathways

Fourier died on May 16, 1830 at the age of 63. ***He had a superb mastery of analytical techniques and notation.*** The integral symbol was his invention. And this power, guided by physical intuition, was the hallmark of his work.

Fourier left behind a legacy, which has been appreciated, only in recent times. Before Fourier, the field of differential equations was a jungle without pathways, where he brought in coherent methods. ***His approach was so original and powerful that a century and a half had to pass for a renaissance of Fourier's theory.***

Augustin Cauchy (1789-1857)

Who developed vital mathematical tools

Cauchy did not master mathematics; he was mastered by it. He was the first mathematician to realise that the greatest material engine of mathematical progress was the printing press. He made abundant use of this medium and literally imprinted his ideas on a generation.

Educated at home

Augustin Cauchy was born in Paris on August 21, 1789. With the onset of the Reign of Terror (1793-94), the family fled to its country house at Aruceil, where they lived for eleven years. Young Cauchy received his education at home: ***religious education from his mother and elementary education from his father.*** He became acquainted with famous scientists — the mathematicians Laplace (1749-1827) and Lagrange (1736-1813) and the chemist Berthollet (1748-1822). His scientific genius was apparent at an early age (as Lagrange forecast in *his warning to Cauchy's father* ***against showing the young boy a mathematical text***).

After eight months of preparation at home, he was admitted to the Ecole Polytechnique, Paris, in 1805. The institution developed as a revolutionary one. Here teaching was linked with research, as the nations finest mathematicians sought to create pure mathematics in discussion with their students and showed them ***how mathematical theory and practice could work in tandem, leading to innovation.***

His career

Upon graduating in 1810, Cauchy was sent to Cherbourg as a military engineer, working on Napoleon's fortifications in preparation for the invasion of England. But he could not stay away from pure mathematics: he pursued research in his spare time. He began his mathematical career in 1811, by solving a problem on the angles of a convex polygon set to him by Lagrange and Fermat's classic problem on polygonal numbers.

In 1813, Cauchy obtained his first teaching position as *repetiteur* and became professor in 1816 at the Ecole Polytechnique. Simultaneously, he was elected to the Academy of Sciences when he was only 27. From 1830, he held professorships in the College de France and the French Academy of Sciences.

Contributions to mathematical analysis

Between 1814 and 1830, he published a series of papers, which laid the foundation for the modern period of rigour in mathematical analysis:

i. Treatise on definite integrals that was to become the basis of the theory of complex functions.
ii. The theory of functions of a complex variable (a variable involving a factor of the $\sqrt{-1}$) which is the principal mathematical tool used in domains ranging from physics to aeronautics.
iii. Propagation of waves at the surface of a liquid; results which are now regarded classics in hydrodynamics.
iv. The founding of the elasticity theory.

In exile with the Jesuits

Cauchy's devoted academic life was suddenly changed by the July Revolution of 1830. Cauchy refused to take the oath of allegiance to the King, which meant that he would lose his Chairs. So he exiled himself, living with the Jesuits at Fribourg working at the University of Turin, and serving as a princely tutor at Prague.

Following his journeys, the steady flow of his papers slowed down. In 1838, when his work in Prague was finished, he returned to Paris and resumed his activity at the Academy.

His publications

In 1826, he founded a private journal, *Exercises de Mathematiques,* which was filled by him with the most improbable choice of topics. Five volumes appeared before he left Paris. In Turin he continued this undertaking, renewed it in Prague and again in Paris, finally reaching a total of ten volumes. He published in other journals and also many textbooks, which have been translated several times.

In this respect, Cauchy was unlike Gauss (1777-1855) who published much less. His works stimulated new investigations much earlier than Gauss' did; and in the range of subject matter, he competes with Gauss.

Cauchy's output was prodigious — about 800 papers; some 500 of these were published late in his life, during his fifties and sixties. In elasticity, there are ***sixteen concepts and theorems named after Cauchy,*** which alone would earn him a place among the greats in science. ***The only person to catalogue his works was Father Jullien, a Jesuit.*** But the catalogue had not been published.

Intolerant of Abel and Galois

In the realm of personal relationships, he showed religious and political bigotry, which made him unpopular. He treated the younger mathematicians, Abel (1802-29) and Galois (1811-1832) in particular, with contempt. As a member of the Academy, ***he delayed by fifteen years, the publication of Abel's masterpiece*** and ***suppressed Galois's epoch-making discoveries.*** By this, Cauchy committed irreparable damage to their careers — Abel died in 1829 and Galois in 1832. Their papers were published posthumously.

A steady stream of mathematical papers trace Cauchy's life: his last communication to the Academy was 18 days before his death on May 23, 1857. He famous textbooks display an exactness unheard of and contain his fundamental work in analysis. His last words were ***"Men die but their works endure".***

Carl Gauss (1777-1855)

Austere mathematician

With Archimedes and Newton, Gauss is ranked as one of the three greatest mathematicians of all time. At an early age, he overturned the theories and methods of 18th century mathematics and opened the way to rigorous analysis. He worked with equal ease in all areas of pure and applied mathematics — number theory, function theory, geodesy, astronomy, mechanics, electromagnetism and optics.

Calculating prodigy

Gauss was born on 30th April 1777, in a poor family of peasant stock in Brunswick (Germany). At an early age he showed signs of unusual talent as a calculating prodigy. It is said that the child, at the age of three, corrected an error in the wage calculation bill prepared by his father who worked as a foreman. ***In his first arithmetic class, he astonished his teacher,*** by instantly writing the answer as 5050, when asked to find the sum of the first 100 integers. Instead of spinning at home to support the family, he was persuaded by one of his friends, Martin Bartels, to continue studies in the gymnasium after school.

Support from Royal family

Impressed by his ability in mathematics and languages, his teachers recommended Gauss, who was then 14 years old, to the Duke of Brunswick for financial assistance. The patronage continued for the next 16 years. In 1792, Gauss joined the Brunswick Collegium. He often discovered important theorems, before coming to know them in his studies.

The years 1796-1800 were most productive: mathematical ideas came to him so fast that he could hardly write them down. In 1799, he earned his doctorate from the University of Helmsteadt, without an oral examination. The subject of his dissertation was a proof of a fundamental theorem of algebra — ***every algebraic equation with complex coefficients has complex solutions. Before he was 25, he was famous as a mathematician and an astronomer.***

Loner

In 1807, he went to Gottingen as Professor at the University and Director of the observatory, which is still functioning. He worked for ***47 years in the post, seldom leaving the city.*** Soon after his death on 23rd February 1855, ***coins were struck in his honour.***

Teaching was his aversion and he could gather only a few students. He was against students taking notes but advised them characteristically, "Don't write, and follow along carefully, instead".

Gauss found no collaborator and worked alone most of his life. Though he was an outstanding practical mathematician, ***he was not inspired in his work by the prospect of practical applications*** but sought truth for its own sake.

Contributions to science: bell-shaped curve

About 1820, Gauss turned his attention to geodesy, the mathematical determination of the shape and size of the earth's surface, to which he devoted considerable time. To increase the accuracy of surveying, he invented the heliotrope, an instrument by which the sun's rays could be reflected in a given direction. It became the standard equipment for large-scale triangulation unit superseded by aerial surveying in the twentieth century.

Gauss showed how probability could be represented by a bell-shaped curve to describe statistically distributed data. This is familiarly known as the Gaussian curve.

System of absolute units

Gauss' most profound work as an investigator of nature is found in his researches on the intensity of the earth's magnetic field, reduced to absolute measure (1832). Thus he founded the system of absolute units, which became indispensable to scientific research. In this, he collaborated with Wilhelm Weber (1804-1891), who was twenty-seven years his junior. Weber's work led to the development of the electrostatic units, while Gauss' work founded the electromagnetic units.

From 1840 the intensity of Gauss's activeness decreased gradually: he suffered from insomnia, bronchitis, heart flutter and such common signs of ageing. Besides, chronic hypochondria and melancholia increasingly plagued him. In the midst of some undated scientific notes from his later years, there suddenly appears the sentence "Death would be preferable to such a life". (*Dictionary of Scientific Biography,* Ch. Scribner's sons, New York)

Collected works

The capacity for work that Gauss displayed was collosal and it is best likened to the contributions of different teams of research workers over many years, in physics, mathematics and geodesy. ***His full stature became known only in the 20th century,*** since he published only about half his recorded research output, in a style so austere that his readers were few.

Many monuments were raised to commemorate Gauss' memory, but the lasting monument is his 'collected works' in 12 volumes. It is an extraordinary edition ***compiled during the period 1863-1933 by ten German Professors,*** each an expert in his area.

Bernhard Riemann (1826-1866)

Who gave geometry a new dimension

Euclidian geometry that is studied at school, postulates the familiar axioms that the angles of a triangle sum up exactly to 180°; parallel lines never meet and so on. The first person, who went beyond Euclid and considered hyperbolic surfaces, was Carl Gauss (1777-1855). It was, in turn, a pupil of Gauss, Bernhard Riemann, who realised the possibility of yet another variation on the theme — a geometry that applied on the closed surface of a sphere. Riemann was one of the most profound and imaginative mathematicians of all time. With only a few published papers, his influence on geometry and algebra was immense.

Shift from theology to mathematics

Born in Breseleu, near Hanover in Germany, on September 17, 1826, Riemann received his early education from his father at home. He joined the local high school in 1840 where he progressed in mathematics beyond the level of even his teachers, mastering calculus and the theory of numbers. In 1846, he enrolled at the University of Gottingen, to study theology and philology.

However, after attending the mathematics lectures, he sought permission from his father (a Lutheran pastor) to shift his field of study to mathematics. He moved in 1847 to the Berlin University for higher studies in mathematics and returned in 1849 to Gottingen. He also attended physics courses under Wilhelm Weber (1804-91).

Field theory

Following studies in experimental physics and natural philosophy, Riemann concluded that ***mathematical theory could connect electricity, magnetism, light and gravity.*** Thus, he suggested the field theory in which spaces surrounding charges of electricity can be mathematically described. During his stay at Gottingen, he had begun to develop original ideas that laid the groundwork for developments in mathematical physics. Riemann earned his doctorate (1851) for his thesis *Foundation for a general theory of function of a complex variable.* He based his treatment on geometrical ideas rather than algebraic calculations alone.

A few years after graduating, Riemann decided on a teaching career at Gottingen. The accepted way at that time in a German University was to seek an unpaid appointment as a 'Privatdozent' whose income solely depended on the fees of the students taking his course. ***The long struggle to survive on such a casual income undermined his health.***

Concepts that underpin geometry

Applicants had to deliver a lecture to the faculty, offering three possible topics for the lecture. It was a tradition that the professor chose one from the first two on the list. Riemann's list was headed by two topics for which he had thoroughly prepared. But the third, an afterthought, dealt with concepts that underpin geometry. Gauss, in his seventies then and still a dominant force at the University, could not resist choosing the third topic, ***on the hypotheses, which form the foundations of geometry.*** It was through Gauss's acumen that ***the splendid idea of this paper was saved for posterity; it became a great classic in the twentieth century.***

A celebrated event

In his ***inaugural lecture (June 10, 1854),*** one of the most celebrated in the history of mathematics, Riemann covered a wide variety of topics, including a workable definition of the curvature of space and how it could be measured, ***the first description of spherical geometry***, and the extension of geometry into many dimensions ***with the aid of algebra***.

However, the most astonishing feature was ***his suggestion that space***

might be curved into a closed ball. This was in hindsight — more than half a century before Einstein came up with the general theory of relativity.

Non-Euclidean geometry

Riemann's idea of curved space and non-Euclidean geometry was introduced to the English-speaking world by William Clifford (1845-1879) *(New Scientist,* Vol. 137, January 2, 1993). After serving three years as a privatdozent, Riemann became an adjunct professor (1857) and then a full professor (1859) succeeding Dirichlet (1805-1859), who had succeeded Gauss four years earlier.

Riemann was beset with overwork and deaths in his family, which led to his faltering health. But he continued producing original papers, which contained rich ideas like the work on partial differential equations. Despite many visits to Italy for recuperation, his strength gradually ebbed. He died when he was only 40, on July 20, 1866 at Selsca, Italy.

His influence on modern mathematics

A measure of his influence is the long list of theorems and concepts that bear his name — Riemann surfaces, Riemann mapping, Riemann integral, Riemann curvature, Riemannian geometry — to mention a few.

His papers

Riemann's evolution was slow. ***As his life was short, his work was not voluminous but was of superb quality.*** His style was conceptual, rather than algorithmic. After more than a century, his papers are still so modern that any mathematician can read them with great delight. ***His courses had inspired a number of books by others on complex function theory.***

EDMUND HALLEY (1656-1743)

THE COMET HUNTER

Halley's calculation of the periodic nature of comets was his most significant contribution to astronomy. The difficulty in determining their motion arose, because a comet could be seen only for a short time and it was only possible to fit a series of curves through observed positions.

Early education

Edmund Halley was born on October 29, 1656, at London, the eldest son of a prosperous landowner and soap-maker for the city of London. Although the latter suffered loss of property in the Great Fire of London (1666), he remained a rich man and spent liberally on his son's education.

Halley was ***tutored at home,*** before being sent to St. Paul's School. At the age of 16, when he entered Queens' College, he was already an accomplished astronomical observer. He left Oxford without a degree.

In 1676, he travelled to the island of St. Helena (South Atlantic), which experienced frequent cloud cover. Still, spending 18 months on the island, Halley managed to determine the position of 350 stars in the southern hemisphere. He returned to England in 1678 and published his results. He was dubbed the ***'Southern Tycho',*** (in honour of the famous Danish Astronomer, Tycho Brahe) by John Falmsteed (1646-1719), the first astronomer royal.

Upon his return to England in 1678, Halley received, by royal mandate, his Oxford degree and was elected FRS. He had obtained by 1684 an inverse square law for the problem of gravitation but since he was unable to deduce from it the planetary motions, he went to Cambridge in August 1684 to discuss it with Newton (1642-1727). He had earlier proved that the orbit of a planetary body is an ellipse.

Friendship with Newton

Realising the significance of what Newton had done in regard to this problem, Halley persuaded the reluctant Newton to develop and publish his work on celestial mechanics. He read the manuscript, contributed important editorial aid and further bore the publication costs. Newton's *Principia* was published in 1687. This association resulted in a lasting friendship; through Newton's influence, Halley was appointed (1696) deputy controller of the mint at Chester. From 1685 to 1693, he also edited the *Philosophical Transactions of the Royal Society* with erudition and academic flair, during the formative years of the journal.

Halley's calculation of the periodic nature of comets was his most significant contribution to astronomy. The difficulty in determining their motion arose, because a comet could be seen for only a short time and it was only possible to fit a series of curves through observed positions. By the mid-seventeenth century, it was generally accepted that the orbital path must be an ellipse, a parabola, or a hyperbola. Newton preferred the parabola, but Halley decided to consider the possibility of an ellipse.

Support to Newtonian gravitation

Using this hypothesis, he made a host of computations; the orbital elements of three bright comets (1531, 1607 and 1682) were so similar that he concluded they were in fact the successive returns of a single body, making ***a periodic appearance of approximately every 75 years.*** Halley predicted the return of this comet in 1758; the comet did reappear on December 25 in that part of the sky he had predicted.

It was not till the occurrence of this celebrated event that the intellectual world of Western Europe took notice of Halley's work. By then, Halley had been dead fifteen years. His hope that posterity would acknowledge his prediction was not misplaced: the object was named Halley's comet, which returned on schedule in 1835, 1910 and 1986. ***His successful prediction gave strong support to Newtonian gravitation.***

Distance of the sun from the earth

His other notable advance in astronomy was in the determination of the distance of the sun from the earth. This was crucial, since a correct evaluation was essential before the size of the planetary system or the distances of the stars could be directly determined. The idea was first sketched in 1663 by James Gregory (1638-1675). Halley first assessed the practicality of this idea, when he observed and timed a transit of Mercury in 1677. By recording the local time at which Mercury appeared to enter the sun's disk and the time at which it left, and then comparing his results with those made at an observing station at a different latitude, the distance of Mercury was obtained. Using Kepler's third law of planetary motion, the distance from the earth to the sun was similarly found.

Ambitious program of study

Halley's other work in positional and navigational astronomy was carried out at Greenwich, after he succeeded John Flamsteed in 1720. With the instruments he added to the observatory, he studied the motion of the moon. This programme was as bold as it was ambitious: for, though he was aged 64, he set about planning observations to cover a complete saros of eighteen years. He adopted this programme because he was sure that once the moon's orbit was known precisely, the problem of determining the longitude at sea would be solved.

The size of the universe

Halley also studied the problem of the size of the universe and the number of stars it contained. He discussed the number of stars to be expected in a given volume of space assuming a given separation between them, and the way in which their brightness would diminish with distance. Here, he anticipated what John Herschel (1792-1871) was to discover a century later that stars of magnitude six were hundred times dimmer than those of magnitude one.

Whatever might be the faults in all this work, ***his methods of approach were new and paved the way for subsequent investigators.***

Stellar astronomy

Halley's most notable achievement in stellar astronomy was his discovery of stellar motion, in the face of the prevailing assumption that stars had to be regarded as fixed. After examining Ptolemy's catalog and those complied 1500 years later, he postulated that stars possessed individual motion. Halley's interest in astronomy was coupled with an equally great interest in pure mathematics. Between 1687 and 1720, he published seven papers ranging from higher geometry to the computation of logarithms and trigonometric functions.

Other contributions

He was also the founder of scientific geophysics. He evolved (1693) an independent novel approach to the question of the age of the earth. From investigations on the rate of evaporation of water, he concluded that salinity of lakes must gradually be increasing and suggested that if the rate of increase could be determined, it should be possible to obtain factual evidence about the earth's age.

Halley possessed great practical sense: he carried out many experiments in diving, designing a diving bell and a diver's helmet. Reports on the colours of light that he observed at various depths were sent to Newton, who incorporated them in his 'optiks'. He also showed ***a lively interest in archaeology,*** publishing a paper (1691) on ***the date and place of Julius Caesar's first landing in Britain,*** using evidence from an eclipse of the moon; and another (1695) on the ancient Syrian city of Palmyra.

Recognition came to Halley early in life with his M.A. (Oxon); but after that, there was a long pause due to the antagonism of Flamsteed. Nevertheless, he was appointed Astronomer Royal in 1720. He was dedicated to astronomy and continued his observations, until a few months before his death on January 14, 1743.

JAMES BRADLEY (1693-1762)

FROM CLERGY TO ASTRONOMY

The introduction of the Kepler telescope for astronomical measurement resulted in increasing refinement in the position of the fixed stars. In view of the annual motion of the earth, astronomers had sought since Copernicus' time of an apparent displacement of nearer fixed starts as compared with more distant ones. This displacement is called the 'parallax of the star'.

After the publication of Copernicus's work in 1543, it became increasingly imperative for astronomers to be able to observe and measure the parallactic displacement of a star because of the earth's annual motion around the sun. By skillfully using improved telescopes, ***Bradley detected the aberration of light, which was the first observational proof for the revolution of the earth around the sun.*** He lived at a time when an astronomer has to be his own technician — repairing, maintaining and making his own optical equipments.

Interest of his uncle

James Bradley was born on March 1693 at Sherbourne, England. He attended the local grammar school and entered the Baliol College, Oxford in 1711. He received the B.A. in 1712 and M.A. in 1717. His father's income was limited; however, he could complete his education through the generous assistance by his uncle, ***James Pound, a clergyman who instilled in the young lad a fondness for astronomy.***

Career as an astronomer

Bradley was ordained in 1719 and the Bishop of Hereford presented him with an additional sinecure living. James Pound, who was also a skilled amateur astronomer, introduced Bradley to ***Edmund Halley*** (1656-1742), the eminent astronomer of that time. The latter was impressed with the erudition and industry of Bradley and ***predicted that Bradley would earn a name in the field.***

In 1721, Bradley was appointed professor of astronomy at the University of Oxford. He resigned his church office and devoted his full time to astronomy. Even as Halley had wished, Bradley succeeded him in 1742 as Astronomer Royal. He held that office with distinction for 20 years.

Bradley was elected FRS in 1718 at the age of 25. Besides he was elected member of the academies of France, Berlin and St. Petersburg. Oxford University awarded him an honorary D.D. in 1742.

In collaboration with his uncle, he derived in 1719 an improved value for the solar parallax from observations of Mars. He attempted the difficult task of measuring the diameters of several stars — Venus, Mars, Jupiter, Saturn and its rings, ***a job that taxed the resources of astronomers about 150 years later.***

His fundamental contributions

On becoming Astronomer Royal, Bradley thoroughly re-equipped the Royal Observatory at Greenwich. He compiled a new catalogue of star positions from observations made during 1750 to 1762. This was published posthumously and involved some 60,000 observations. Bessel's catalogue of 1818, with 3000 star positions, was largely based on Bradley's work.

Bradley, in addition, took painstaking interest in the accurate measurement of time. He worked out a correction for the higher temperatures in Jamaica and deduced a slowing of the clock by 1 minute 59 seconds per day, which agreed with Newton's theory of the relation between latitude and gravity.

Bradley's celebrated *discovery of the aberration of light* is a good example of the way in which his patience, industry and clarity of perception was crowned by success, when earlier attempts by Robert Hooke (1669) and Samuel Molyneux (1725) were unsuccessful.

It is said that the explanation for this phenomenon came to Bradley as he sailed on the Thames, observing how the wind vane on the mast shifted position with the varying motion of the boat, even though the wind had not changed direction. He concluded that ***the apparent stellar shift was brought about by the aberration of light,*** which was a result of the finite speed of light and the forward motion of the earth in its orbit.

This was a fundamental contribution, memorable for the fact that Bradley gave precisely the modern value for the constant of aberration, about 20.5 s (presently 20.47 s). Out of his work on aberration, Bradley *discovered nutation;* the oscillation of the earth's axis caused by the changing direction of the gravitational pull of the moon on the equatorial bulge.

Work at the Greenwich Laboratory

Here Bradley spent his entire professional life; it was founded to assist navigation — to increase the safety of ships on the ocean, by prescribing better methods of finding the longitude at sea. He recognised the importance for navigation of magnetic observations and accordingly added magnetic instruments.

In 1759 and 1760, Bradley compared Mayer's lunar tables for finding the longitude at sea with many more observations and worked out corrections for them by intricate and tedious calculations. Bradley was a very abstemious person. ***He was able to endure long hours of observation and laborious calculations.***

In the last stage of his life, partly through overwork, his health gradually deteriorated. By 1761 he became unfit for regular work, as he was obsessed by the fear that his brain was giving way. He passed away on July 13, 1762 at Chalford. Bradley's work at the Greenwich Laboratory brought him pre-eminence among both astronomers both in England and on the Continent.

Without his two great discoveries and his work on refraction, it is difficult to see how later progress by other astronomers in the determination of star positions, distances and motions would have been possible.

(*Dictionary of Scientific Biography,* Vol II)

WILLIAM HERSCHEL (1738-1822)

FATHER OF STELLAR ASTRONOMY

Careers of many scientists have been shaped by political events, but in the case of the Herschel family, they appear almost genetically determined. This family, by their indefatigable studies of the stars and nebulae, left their indelible imprint on British science: William's sensational discovery of the planet Uranus (1871); the discovery by his sister Caroline (1750-1848) of eight comets and three nebulae; his only son John (1792-1871) taking up astronomy out of a sense of filial devotion and completing his father's work.

Teaching music

William Herschel was born on November 15, 1738 in Hannover, Germany and following his father's profession, he played in the band of the Hannovarian Guards. After the French occupation of Hannover, he escaped in 1757 to England where he earned a living by copying music, eventually becoming (1766) a music teacher and an organist at a fashionable chapel in Bath.

By his own intellectual curiosity, William was led from the practice to the theory of music, influenced by Smith's 'Harmonics'. ***Smith's other book 'Opticks' introduced him to optics and astronomy*** and whetted his appetite for viewing the night sky.

Pursuit of astronomy

In 1772, William fetched his sister Caroline from Germany to live with him in Bath. She began a career as a singer but as William became obsessed with astronomy, she carried out duties as his faithful assistant through his entire career. News of this extraordinary household began to spread in scientific circles.

By 1779, William had made two telescopic surveys of the heavens. During his third and most complete survey of the night sky (1871) he came upon an object, that, he defined was not an ordinary star. It proved to be the planet Uranus, the first planet to be discovered since the dawn of history. William became famous almost overnight. His friend Dr. William Watson helped him to get in 1782 an annual pension of £200 from King George III. He could thus give up music and devote himself exclusively to the pursuit of astronomy. At this time William was appointed as an astronomer and his duty was to show the heavens to the Royal family from time to time.

Grinding and polishing large mirrors

Although William was 43 years old when he became a professional astronomer, he worked night after night to develop theories of the 'construction of the heavens'. Newton, Bentley, Halley and others had discussed questions on the number of stars and the nature of the nebulae. But the necessary basis of evidence was totally lacking. It was William's achievement to assemble a mass of observational evidence and make it the basis for bold theorising. He seemed to have realised quickly that in order to investigate very distant objects, he would need telescopes with considerable light-gathering power. He soon took up the construction of grinding and polishing large mirrors from metal disks of copper, tin and antimony in various proportions.

Study of nebulae

It took four years of labour for Herschel and his team to build a mirror of 4 feet diameter and 40-foot focal length. The monster telescope, when completed in 1789, became one of the wonders of the 18th century and a visible testimony to his mechanical ingenuity. William's interest in the nature of nebulae, which appear as luminous patches in the sky, developed in 1781-82.

He found that his most powerful telescope could resolve into stars several nebulae that appeared 'milky' to less well equipped observers. He argued convincingly (1785) that all nebulae were formed of stars and that there was no need to postulate the existence of a mysterious fluid to explain the observed facts.

Theory of the evolution of stars

In three important papers (1784, 1785, 1789) William developed a cosmogony — a theory concerning the origin of the universe: the stars originally were scattered throughout infinite space, in which attractive forces gradually organised them into even more fragmented and tightly packed clusters. Thus William made change in time, or evolution, a fundamental explanatory concept in astronomy.

In October 1783, ***he embarked on an intensive 20 years program of sweeping for nebulae and eventually raised the total of those known to 2500.*** In 1787, the Herschels moved to Old Windsor and the following year to nearby Slough, where William spent the rest of his life (August 25, 1822). Night after night, he observed the sky in the company of Caroline. Often in the daytime, Caroline would summarise the results of their work, while he directed the construction of telescopes, many of which he sold to supplement his income.

William's labours through systematic sweeps for nebulae (1783-1802) resulted in three catalogs listing 2500 nebulae that he substituted for the 100 or so milky patches previously identified. He also catalogued 848 double stars, pairs of stars that appear close together in space and listed the comparative brightness of stars. He published 70 papers covering investigations of the planets and other members of the solar system, amongst which there was an announcement, in 1800 of the discovery of infrared rays.

William, with his total dedication to his work, seemed not to have contemplated marriage till 1788, when he was nearly 50. In the same year he married Mary, the widow of his friend John Pitt. The displacement in her brother's affections came as a blow to Caroline, but she was won over by Mary's kindness.

Honours

The Royal Society, London awarded William the Copley Medal and elected him a Fellow. He was knighted in 1816 for his contributions to British science.

Recognition of Caroline's work

William could achieve so much in a field, which he entered only in middle life because of ***the selfless support of his sister*** Caroline. She carried out the extensive routine calculations, prepared catalogs, ground and polished mirrors, besides managing the household. ***In 1787 her work was recognised*** by a pension of £50 per annum from the King. Between 1786 and 1797, she discovered no fewer than eight comets, thus earning a reputation as an astronomer in her own right. She returned to Hannover in 1822, where she spent the rest of her life, vigorous and alert to the end (March 16, 1850). In her old age, she received many honours.

As the years passed, the years of unremitting labours through twenty years took their toll on William's health. Seeing the failing health of his father, John left Cambridge in 1816 and became his assistant. In this way, the unrivalled experience of William Herschel was transmitted to the next generation.

(*Dictionary of Scientific Biography,* Vol. VI.)

JOHN HERSCHEL (1792-1871)

WORTHY SON OF AN ILLUSTRIOUS FATHER

He left Cambridge for good to join his father out of a sense of filial devotion; the latter, at 78, was anxious to see his astronomical work continued. But John held no paid post between 1816 and 1850.

Education at Cambridge

John Herschel, a scion of the celebrated Herschel family, was born on March 7, 1792 at Slough, England. When he was eight, he was sent to school at nearby Eton; but his mother, seeing him involved in a fight, took him away to be educated privately.

In 1809, John entered the Cambridge University, where he gained all the advantages of education and society that his father had missed. Here he made friends with men of great ability: George Peacock (1791-1858), the mathematician; Charles Babbage, (1792-1871), the innovator of digital computation; and William Whewell, the natural philosopher. While still undergraduates, ***the three resolved "to do their best to leave the world wiser than they found it" and founded in 1812 the Analytical Society,*** in order to introduce continental methods of mathematical analysis into English curriculum. ***They did so by replacing the cumbersome symbolism of Newton with the more powerful notations by the German mathematician Leibniz (1646-1716).***

John Herschel took the tripos oral examination in 1813. He was named senior wrangler and Smith's prizeman, with Peacock next. Babbage withdrew, because he could not compete with Herschel. His paper on the application of Cotes theorem brought him FRS (1813) at the age of 21. He missed the election to the post of professor of chemistry at Cambridge by only one vote. He was an expert chemist, too.

Choice of career

So John vacillated in the choice of his career. His father favoured the church, for "a clergy is in possession of dignified leisure to cultivate his mind". But John preferred law, enrolling in 1814 at Lincoln's Inn where he used much of his time to develop his friendship with William Wollaston (1766-1828) who discovered the dark lines in the solar spectrum and James South, a wealthy amateur astronomer. Both men exerted strong influences on his later scientific work.

Continuing his father's work

John accepted a minor teaching post at St. John's College, but finding this work incogenial, he left Cambridge for good to join his father out of a sense of filial devotion; the latter, at seventy-eight, was anxious to see his astronomical work continued but felt his own powers declining. How this was financially possible was puzzling, as ***John held no paid post between 1816 and 1850.*** He brought up a large family of twelve children and spent large sums on his expedition to Africa. His books sold well but could have hardly provided his livelihood.

John's first major task was the re-observation of the double stars that were catalogued by his father. In James South, he found a collaborator who could provide two excellent refracting telescopes needed for this work. The catalogues they compiled between 1821 and 1823, which were published in the Philosophical Transactions (1824), won them the Gold Medal of the Royal Astronomical Society and the Laalande Prize of the Academy of Sciences, Paris.

His marriage

John's studies of double stars led him to the very important ***problem of measuring the parallax of fixed stars,*** which is apparent in angular position of a given star caused by the earth's revolution around the sun. By the time he was nearing forty, he had become famous in his field. He might have remained a solitary bachelor but for the interest of his friend Grahame. The latter arranged his marriage (1829) with Margaret who was eighteen years younger than John. Margaret followed him even to the wild of Africa and ran a household of seldom less than twenty people, when she was still in her early twenties.

Expedition to the Cape of Good Hope

John's sense of obligation to complete his father's work led him to plan an expedition to the Southern Hemisphere to survey the skies not visible in England. In November 1833, John with his family set sail for the Cape of Good Hope with a large reflecting telescope for observing faint nebulae. It had three interchangeable mirrors, one made by William, one by John under his father's directions and the third by John alone. He spent four years in intense scientific activity.

When John sailed for home in March 1838, he had recorded the locations of 68,948 stars, described many of the details of the Great Nebula in the constellation of Orion, and observed Halley's comets and the satellites of Saturn.

In his journey, John made many physical and meteorological experiments: ***he invented a device called an actionmeter,*** which consisted of a large bulb thermometer containing a dark liquid. ***Comparing the rate of rise of this liquid in the sun and in the shade, he derived a numerical measure of the solar energy***. He introduced numerical measurements of stellar photometry with the astrometer he invented, by which he could compare the brightness of stars with an image of the full moon.

A labour of love, John proudly sent his aunt Caroline (1750-1848) in 1847 the massive 'Results of Astronomical Observations made during 1834-38 at the Cape of Good Hope'.

Between 1821 and 1824, he undertook a number of European tours: he met all the leading scientists of his day — Arago, Laplace, Blot, Humboldt, Gay-Lussac, Poisson, Fourier and Gauss.

Publication of a text for the layman

John founded the Astronomical Society, placating the hostility of the senior Royal Society and was elected president three times (1827, 1839, and 1947). He was made a Baronet on his return from Africa. He laboured, taking on additional commitments — Master of the Mint accepting the public office and composing the 'Outlines of Astronomy' for the educated layman (1849). This highly successful text went through many editions including Arabic and Chinese translations.

Other achievements

John published important studies of his own on the new subject of photochemistry invented by the French photographer Louis Daguerre and succeeded in preparing a photograph of a solar spectrum in its colours. He discovered the ***'Herschel effect',*** the quenching effect of light of a longer wavelength on a photosensitive surface afterwards exposed to a shorter wavelength.

As a member of the committee choosing scientific instruments for the Great Exhibition of 1851, he was forced to live in London away from his family. His health deteriorated and in 1854, he suffered a nervous breakdown and was incapacitated for several weeks. So in 1856, he resigned his post at the mint but still had the verve to produce a consolidated catalogue of 5079 nebulae, and star clusters and to do much of the work for the posthumous catalog of 10,300 double stars. Besides, John found ***time to translate into English some of the works of Schiller, Dante's Inferno*** and, towards the end of his life, the Iliad.

When John Herschel died on May 11, 1871, he was mourned by the whole nation not merely as an academician and scientist but also one of the last of the universalists (*The New Enyclopaedia Brittanica,* Vol. 5).

HENRI POINCARE (1854-1912)

'MONSTER' OF MATHEMATICS

One new idea of his was destined to revolutionise physics at the end of the 19th century. Poincare's remark on a possible connection between X-rays and the phenomenon of phosphoresence was the starting point of the discovery of radioactivity.

Extraordinary capability

Poincare was born at Nancy (France), on April 29, 1854 into an upper middle class family. His mathematical ability, while still a student in the 'lycee' was conspicuous. He also excelled in written composition in the school.

Poincare was ambidexterous and could not see clearly the mathematical symbols written on the blackboard. But he overcame this handicap wltlh his phenomenal memory, by which he could visualise what he heard. His professor at Nancy nicknamed him as a 'monster of mathematics'. ***He could perform complex mathematical calculations mentally, a capability which he retained throughout his life.***

In 1872, Poincare entered the Ecole Polytechnique in Paris at the top of his class. He graduated in 1875, winning honours in mathematics. Then he attended courses in engineering at the Ecoles des Mines and worked for a short period as an engineer. He obtained his doctorate in mathematics in 1879, after which he held a post in mathematical analysis at the University of Caen.

Career at the Sorbonne

Poincare joined the University of Paris (Sorbonne), where he served until his death at the age of 58 on July 17, 1912. He gave himself up to his duties with his characteristic thoroughness and energy. ***Many of his lectures were printed shortly after they were delivered*** at the Sorbonne. His prestige and influence thus increased in French science.

Honours

Poincare was elected, at the age of 33, to the Academy of Sciences, of which he became President in 1906. In 1908, he became a member of the Academy of France the highest honour in his country. Poineare wrote prolifically, ***almost 500 papers,*** on a wide range of topics in celestial mechanics, ***in pure and applied mathematics and in algebraic geometry.***

Major contributions

i. *Celestial mechanics*: Poincare took up the rigorous treatment of the subject in contrast with the semi-empirical computations that had been prevalent before him. He made substantial contributions to the theory of equilibrium of rotating fluid masses, and the theory of the solar system; particularly the classical three-body problem, for example, the system involving the sun, moon and earth.

This was part of the 'problem of n-bodies' set for a prize by King Oscar II of Sweden, which was awarded to Poincare's paper. This paper ranked as one of his masterpieces; it dealt with the periodic solutions of the three-body problem when the masses of two of the bodies are very small in relation to the mass of the third (as in the solar system).

ii. *Differential equations*: The period from 1880 to 1883 is considered to be his prodigious period of creativity, when he formulated the qualitative theory of differential equations. It is one of the few examples of a mathematical theory that sprang apparently from nowhere and immediately reached perfection in Poincare's hands.

iii. *Function theory:* Before he was thirty, he became famous by his discovery of the 'automorphic function', the function that is invariant under a group of transformations characterised by the ratio of linear terms. He was the first to generalise the concept of residue for multiple integrals of functions of complex variables.

When Poincare started his research, it is said ***he was hardly aware of the mathematical literature on the subject;*** he hardly knew anything on modular functions beyond the work of Hermite, his teacher. He certainly had never read Riemann (1826-66) and had not even heard of the 'Dirichlet principle', which he was to use in such an imaginative way a few years later.

Poincare was open-minded regarding new mathematical ideas. He was probably the first mathematician to use Cantor's theory (1854-1918) of sets in analysis; he had met concepts such as perfect non-denuse sets in his work on automorphic functions.

His paper on number theory (1901) was most influential and was ***the first paper on what is now called algebraic geometry*** over 'a field of algebraic numbers'.

iv. *Mathematical physics:* For more than twenty years, Poincare lectured at the Sorbonne on theoretical physics, with a preferred accent on the theories of light and electromagnetic waves. ***One new idea of his was destined to revolutionise physics at the end of the nineteenth century.*** His remark on the possible connection between X-rays and the phenomenon of phosphorescence was the starting point of Becquerel's famous experiments that led to the discovery of radioactivity.

Poincare formulated (1906) in a paper on the dynamics of the electron, independent of Einstein, many of the results of the special theory of relativity.

v. *Algebraic topology:* In 1894, he commenced a series of six remarkable papers, which continued over a decade, on the modern methods of algebaric topology. The machinery of what is now ***'simplicial homology' is entirely a creation of Poincare.*** In two of the papers of this series on analysis situs (continuity), he determined the fundamental group of surfaces defined by an equation of the type $z^2 = F(x, y)$, thus paving the way for the generalisation later given by Lefschetz and Hodge.

vi. *On mathematical discovery:* As his international prominence rose, ***Poincare turned his superb litrary gifts to the challenge of speaking or writing on mathematics for the general public.*** His vivid style and clarity of mind earned him the reputation as the best communicator of science to the layman.

His well-known descriptions of ***the process of mathematical discovery*** remain unsurpassed. In his view, ***sudden illumination following long subconscious work***, is a prelude to mathematical discovery. This has been corroborated by many mathematicians.

GEORGE CANTOR (1854-1918)

WHO FOUNDED THE SET THEORY

The dual concepts of infinitesimals and infinities have always been the nucleus of crises from the first foundational crisis that some have associated with the discovery of irrational numbers to the debates that had been waged between the descendants of Kronecker and Brouwer (intutionists) on the one hand and of Cantoi and Hilbert (formalists) on the other.

George Cantor was the first to speak ***the language of modern mathematics*** and the first to deal rigorously with the transfinite entities that are one of the central notions of modern mathematics.

Early education

Cantor was born on March 3, 1845 in St. Petersburg (Russia). He hailed from a family which included a renowned violin virtuosi, because of which, he later voiced regrets that his father had not let him become a violinist. He entered in 1860, the gymnasium in Weisbaden; at 15 his mathematical talent was already evident. In 1862, he enrolled for higher studies in the university at Zurich, shifted to Berlin in 1863, after the sudden death of his father.

At Berlin his teachers were Weierstrass, Kummer and Kronecker. Weierstrass (1815-1897) had a great influence on Cantor, as the former's lectures on analysis gave Cantor a firm and precise foundation. Cantor learnt much from Kronecker (1823-1891), the specialist on the theory of numbers. He received in 1867 his doctorate.

Cantor joined in 1869, the University of Halle as a 'privatedozent' (unpaid lecturer), soon becoming an assistant professor, and in 1879, a professor. This position at Halle was so poorly paid that, only with the financial support he derived from an inheritance from his father, could he spend his whole career here.

His battle with other mathematicians

While pursuing a problem on trigonometric series, Cantor was led to investigate the nature of infinite series of numbers. He developed the theory of sets and transfinite sets in a collection of papers from 1874 to 1897. They were published in the then new *Acta Mathematica.* His creation of transfinite entities demanded great courage, as he had to overcome the prejudices of many generations of mathematicians. The chief among Cantor's opponents was Kronecker who, using his prestige, not only prejudiced others against Cantor's ideas from his outpost in Halle but thwarted Cantor's every attempt to improve his standing by winning a better endowed and more prestigious professorship in Berlin.

Naturally, his health was impaired by the dual strain of trying to solve the problem of 'continuum hypothesis' and of trying to withstand the opposition of other eminent mathematicians to his pioneering work. In the end, Cantor prevailed, although ***the battle nearly cost him his sanity. Since 1884, he had suffered frequent spells of deep depression and had to often remain in a sanatorium.*** It was sad that his end came in Halle University's psychiatric clinic, on January 6, 1918.

A new province for mathematics

In 1873, an exchange of letters with his colleague Dadekind in Brunswick channeled all of Cantor's subsequent work in a new direction. Cantor wrote that it was possible to 'count' the set to rational numbers, that is, to pull them into a one-to-one correspondence with the set of natural numbers; but he wondered whether such one-to-one mapping was not also possible for the set of real numbers. Cantor continued with his efforts and by ***December 7, 1873,*** he was able to write to Dadekind that he had succeeded in proving that the 'aggregate' of real numbers was uncountable. That date ***can probably be regarded as the day on which set theory was born.***

Cantor regarded his theory as an expansion of classical number theory. He introduced 'transfinite' numbers (cardinal numbers, ordinal numbers) and developed an arithmetic for them. With these numbers, explained in terms of transfinite sets, Cantor indeed opened up a new province for mathematics.

New areas of enquiry

Cantor's unique contribution was that his special way of asking questions unearthed vast new areas of enquiry, in which the problems were solved partly by him and partly by his successors. He provided the first satisfactory definition of the word 'continuum'. A continuous perfect set he called a continuum, thereby turning that concept, very vague until then, into a useful mathematical tool.

A much greater stir was caused in 1895 by the antinomy involved in 'the set, of all sets which do not contain themselves as an element'. ***It was chiefly Hilbert (1862-1943) who, looking for a way out of the impasse, proposed a strict formalisation of set theory.*** Thus he brought about the dawning of the modern age of mathmatics, based on Cantor's firm foundation.

Integrating set theory into metaphysics

By his bold advance into the realm of the infinite, ***Cantor ignited twentieth-century research.*** He was axiomaticist, his way of thinking belonged more to the classical epoch. He is considered the last great Platonist among mathematicians. He intergrated his theory of sets into metaphysics. He also sought to tie it with theology, which used metaphysics as its scientific tool.

Cantor was convinced that the actual infinite really existed both concretely and abstractly. "While possibly I am the very first in history to take this position so explicitly, I know for sure that I shall not be the last", he wrote.

In 1908, the great mathematician Henri Poincare (1854-1912) remarked that "later generations would regard Cantor's set theory as a disease from which one has recovered". Poincare was wrong. Today, ***the power and use of set theory to illuminate and unify areas of mathematics*** is widely recognised.

DAVID HILBERT (1862-1943)

A LIFE IN MATHEMATICS

> What a great charm he wielded on his students: "The sweet flute of the Pied Piper Hilbert, seducing so many rats to follow him into the deep river of mathematics."

Effortless learning

David Hilbert came to Gottingen University in March 1895, exactly eighty eight years after Gauss (1777-1855). It was not immediately apparent to the students that another great mathematician had joined the Gottingen tradition. If he was engrossed in integral equations, integral equations seemed everything to him; dropping a subject meant dropping it for good and turning to something else.

Born on January 23, 1862 at Koenigsberg (Germany), he attended the Friedricks-kolleg from 1870; he completed his high school in the Wilhelm Gymnasium. Mathematics appealed to him as it was 'bequem' (effortless) and required no memorising. He was happpy here, as the teachers recognised and encouraged his originality; and in later years, he would often recall them with affection.

At Koenigsberg (1880-1894)

In 1880, Hilbert joined the University of Koenigsberg and completed the eight semesters required for a doctorate. He was considering for his dissertation the investigation of continued fractions but his ***Professor Lindemann suggested instead a problem in the theory of algebraic invariants.*** This was a turning point: though the problem was attractive, considered a very modern subject, its roots lay in the seventeenth century. On Descartes's map of the plane, any point on the plane is equivalent to a pair of real numbers *x*, *y*; geometric figures can thus be formalised by algebraic equations and conversely, algebraic equations can be graphed as geometric figures. His professor highlighted a number of concepts and relations that remained to be clarified.

A new way of learning

After receiving his Ph.D. in 1885, he visited the universities at Leipzig and Paris. He remained at Koenisberg as a privatdozent during 1886-92 and succeeded Lindemann as professor in 1893. During this period, he came under the influence of two mature mathematicians, Herman Minkowski and Adolf Hurwitz. With them he exchanged his thoughts and scientific plans during evening walks. Hurwitz, with his vast well-founded and well-ordered knowledge, was always the leader. Hilbert found this way of learning was infinitely preferable to poring over dusty books in the library.

Early in December 1894. Hilbert received ***an invitation from Felix Klein*** (1849-1925), the legendary figure in mathematical circles: "You are the man whom I need as my scientific complement because of the direction of your work and the power of your mathematical thinking. ***I am counting that you will give a new inner strength to the mathematical school*** here, which has grown continuously and that perhaps you will even exercise a rejuvenating effect upon me..."

Professor at Gottingen (1895-1930)

Apparently Hilbert had already absorbed a feeling for Gottingen, the university of Gauss, Dirichelet and Riemann. So Hilbert could easily make up his mind to join Gottingen University, where he served from 1895 till his official retirement in 1930.

The first year he lectured on determinants and elliptic functions; besides, he conducted every week a joint seminar on real functions with Felix Klein.

For Max Born (1882-1970) and other students, the seminar meetings were stimulating hours.

His fame

Hilbert's scientific activity during his active career spanned six periods:

i. 1885-1893 Theory of invariants
ii. 1893-1898 Theory of algebraic numbers
iii. 1898-1902 Foundations of geometry
iv. 1902-1910 Integral equations
v. 1910-1922 Theoretical physics
vi. 1912-1930 Foundations of mathematics

Hilbert's fame rests on ***a list of 23 research problems he enunciated in 1900*** at the International Mathematical Congress in Paris. These have been solved in the next forty years: ***a mathematician who had solved one of them passed on to the Honours class of Mathematicians*** (*Bulletin of the American Mathematical Society,* H. Weyl, Vol. 50, 1994).

Advices in pedagogy

His lectures — so different from Klein's, which were recognised as classics — were ***full of most beautiful insights into mathematical thinking.*** Sometimes, he would develop one of his own ideas spontaneously in front of the class. Thus he touched and enlivened the students.

Hilbert practised his own theory in pedagogy. He did not have a very high opinion of the ordinary student and believed that nothing much was absorbed until it was heard several times. "Repeat five times!", ***"Keep computations to the lowest level", "Begin with the simplest examples"*** — these were the weighty advices which he himself followed and gave to his assistants. He presented important ideas in vivid forms, looking always for contrasts to make them striking and memorable.

Hilbert went further and ***taught his students how to frame a problem and solve it.*** "He would take a good part of the hour to explain a question, then the formal proof, when it followed, would appear to us so natural that we wondered why we had not thought of it ourselves", recalled one of his students.

His research work

In the summer of 1899, Hilbert turned to an old and celebrated problem known as the, 'Dirichlet Principle', which involved all the great mathematicians of Gottingen. He proved in 1904 the Dirichlet principle by brute force technique and perplexed the mathematical world by salvaging it.

Hilbert's work in integral equations (1909) led directly to 20th century research in functional analysis, the branch of mathematics in which functions are studied collectively. His work also established the basis for infinite dimensional space, called ***Hilbert space,*** a concept which is useful in mathematical analysis and quantum mechanics.

The work on integral equations (published as a book in 1912) naturally brought him to the boundary of mathematics and physics. In particular, his asymptotic solution of Maxwell–Boltzmann's fundamental equation of the second order, clearly separated the two layers of phenomenological physical laws to which the theory leads.

His view on physics

But as a mathematician, he was disturbed by a certain lack of order in the triumphs of the physicists. Physics, he asserted, was too difficult to be left to physicists! This might seem an arrogant remark, but ***one Nobel Prize winner later unravelled the point:*** "What he expressed thus was something completely genuine; the respect for the difficulty of the problems which are posed in this field of pure thought, recognised only by one who has actually put all his intellectual power to overcoming such problems". (*Hilbert,* Constance Reid, George Allen and Unwin Ltd. London, 1976)

Hilbert died at Gottingen on February 14, 1943. There was scarcely a mathematician in the world, whose work did not derive in some way from that of Hilbert.

NORBERT WIENER (1894-1964)

FATHER OF CYBERNETICS

Wiener's understanding of modern mathematics began in 1918, when he came across works on integration and functionals. One of his colleagues suggested to him to work on integration in function spaces which put him on the track leading to his greatest achievments.

Tried different subjects at college!

Norbert Wiener, ***a child prodigy*** who could read and write at the age of 3, was born on November 26, 1894, in Columbia, U.S.A. His father, professor of literature, at the Harvard University, was a man of unusual intellectual ability and personally supervised his early education about which Norbert recalls vividly... "I learned my science for myself; my father introduced me to languages, both ancient and modern, and to mathematics. All these subjects held a certain interest for me, but no casual interest could satisfy my father's demands for precise and ready knowledge, which were severe and painful". Wiener graduated from Tufts College at 14, with a degree in mathematics.

Then, he enrolled in Harvard University ***to study zoology*** but had to quit, as he was inept at laboratory work. He next tried ***to study philosophy*** at Cornell. Finally, on the advice of his father, who dominated his life well into manhood, he shifted to Harvard, from where he received a Ph.D degree in 1913 with ***a dissertation on the boundary between philosophy and mathematics.***

A delight

On a Harvard fellowship, Wiener then studied under Bertrand Russell (1872-1970) at Cambridge and David Hilbert (1862-1943) at Gottingen. His thinking was moulded by the English mathematical theorist, G.H. Hardy (1877-1947): ***"His course was a delight to me.*** Hardy led me through the complicated logic of higher mathematics with such clarity ***that resolved many difficulties and gave me a real sense of what is necessary for a mathematical proof.*** He introduced me to the Lebesgue integral, which was to lead directly to the main achievements of my early career". (*I am a Mathematician,* N. Wiener, MIT Press, 1956)

Back in the U.S. in 1915, ***Wiener tried a variety of jobs*** — a teacher, a writer for an encyclopedia, an apprentice engineer, a journalist of sorts but was not successful in any of them.

Entry into MIT

Finally, in 1919, Wiener got a position as an instructor in the mathematics department of the Massachusetts Institute of Technology. It turned out however, to be the right move for him. He served on the faculty for 35 years till his retirement in 1954 and by his original contributions ***raised its stature to a great centre of learning in science and technology.***

Wiener's understanding of modern mathematics began in 1918 when he came across works on integration and functionals. One of his colleagues suggested to him to work on integration in function spaces, which put him on the track leading to his greatest achievements. He was not merely satisfied with a general integration theory but looked for physical embodiments to test the theory. The first one he tried, turbulence, was a failure; but the next, Brownian motion, was a success. He then took up the study of stochastic processes: one such problem, in which Wiener became involved during World War II, is gun fire control.

Publication of the 'yellow peril'

The problem was to build into the control system of the gun, a suitable mechanical device to aim the gun automatically. It was, in effect, made up of two parts: ***a mathematical part,*** which consisted of predicting the future position of an airplane from its observed past positions, and ***an engineering part,*** which consisted of realising the mathematical solution in the form of an actual physical device. The brilliant work, which involved fundamental concepts, was published in 1942 as a classified wartime report.

It was bound in a yellow paper cover and, ***because of its difficult mathematics; it came to be known among engineers as the 'yellow peril'***. Its full title is 'Extrapolation, interpolation and smoothing of stationary time series with engineering applications', and it was republished in 1949 by MIT Press, Cambridge.

Building the control system

However clear in a conceptual way, the building of an actual device was to Wiener, "a quite simple assembly of electric inductances, resistances and capacitors acting on a small electric motor", ***there were few engineers at that time who were able to grasp Wiener's solution and realise it in the form of a physical device.***

The credit for working out Wiener's ideas goes to his outstanding student Levinson, who describes his contact with Wiener as follows: "He was a most stimulating teacher. ***He would actually carry out his research at the black board***.... I found a gap in a proof of Paley–Wiener equation and proved a lemma to set it right. Wiener sat down at his typewriter, typed my lemma, affixed my name and sent it off to a journal. A prominent professor does not often act as a secretary for a young student!" (*Proc. IEEE,* Vol. 70, September 1983.)

Formulation of cybernetics

The union of the software and hardware research interests led to Weiner's discovery and formulation of the *science of cybernetics.* His classic work, *Cybernetics or the Control and Communication in the animal and the machine* (1948) exerted great influence in the area of control and communication engineering. It has contributed to popularising ***a way of thinking in communication theory by terms such as feedback, information, control, input, output, stability, prediction and filtering.*** It is said that it has also contributed to spreading mistaken ideas of what mathematics really means. This work earned the greater part of his public renown.

Wiener was egocentric and worried about the quality of his work and his standing as a mathematician; yet he was also extremely generous, with a strong sense of social and moral responsibility. He was a vegetarian, something of a puritan. He travelled a great deal. He died on March 18, 1964, during a trip to Sweden.

Honours

A few weeks before his death, Wiener received ***the National Medal of Science.*** His last book *God and Golem Inc.* won (1965) the American National Book Award in the category of science, religion and philosophy.

Srinivasa Ramanujan (1887-1920)

The greatest mathematician of India

In a lone mighty sweep, Ramanujan had succeeded in recreating in his field, through his own unaided powers, a rich half century of European mathematics. He occupies, along with Euler, Gauss and Jacobi, an immortal place in mathematics.

Celebrity in the school

Ramanujan was born on December 22, 1887, in Erode (South India) as the eldest son in a family of six children. He entered in November 1892, the Town High School at Kumbakonam as a half-fee scholarship holder and passed the Matriculation examination in 1904. In the school, he came something of a minor celebrity, walking off with merit certificates and prizes for academic brilliance. ***This school nourished him for six years,*** bringing him as close as he would ever come to a satisfying academic experience.

Early evidence

When he was in the seventh standard, he gave clear evidence of his mathematical gifts: he could reel off the square root of a natural number to the specified number of places; he could point to the indeterminate nature of zero divided by zero. Ramanujan's mother — the family being close to penury — took in college students as boarders: noticing Ramanujan's interest in mathematics, they brought him textbooks from the college library; Loney's 'Trigonometry' was one such treasure which he mastered.

Ramanujan came to understand ***trigonometric functions not merely as the ratios of the sides in a right-angle triangle but as concepts involving infinite series.*** He would rattle off the values of transcendental numbers such as *pi* and *e* to any number of decimal places. Classmates, two years his senior, would seek his assistance with problems they thought difficult, only to watch him solve them at a glance.

Similarly, ***Carr's 'Synopsis'*** came into his hands through his friends, before he joined the Government College, Kumbakonam, which had earned the moniker 'the Cambridge of South India'. In essence, the book was a compilation of about 500 equations in algebra, calculus, trigonometry and analytical geometry with abridged demonstrations of the propositions. "The book is not in any sense a great one" some mathematician remarked later, "but ***Ramanujan has made it famous".*** Working on this book gave him hours of delight; verification of the formulae, justification and finally entry into mathematical thought.

Formal education

With the high standing in his school record, he joined in 1904 the F.A. course. He failed in English composition and lost his scholarship. He could not bear to witness the grinding poverty of his parents, yet he did not give up his passion for mathematics. Unable to endure the situation, in August 1905, he ran away from home. It was the first of numerous such occasions, about which little became known. On, July 14, 1909 Ramanujan married Janaki, then nine years of age. Ramanujan joined Pachaiyappa's College in Madras in the F.A. class again. He could take the three-hour mathematics paper and finish it in thirty minutes, which got him nowhere.

In December 1906, he appeared for the F.A. examination and failed. The following year, he took it again and failed. This marked the end of his formal education.

"In proving one formula, as he worked through Carr's synopsis, he discovered many others, and he began the practice of compiling a notebook", wrote E.H. Neville, an English mathematician who later befriended

Ramanujan. Such was the likely genesis of ***his notebooks, which won him an immortal place along with Euler, Gauss and Jacobi.***

Evaluation of π (pi)

In the mid-seventeenth century, series approximations were used to evaluate *pi*. Many rested on modular equations going back to the work of Legendre in 1825. Leibniz's series though looking lovely, demanded no fewer than 500 terms to converge to *pi*, but yeilded only three-decimal place accuracy. In one of Ramanujan's series, the very first term gave *pi* to eight places. Many years later, this would provide ***the basis for the fastest known computer algorithm to determine pi.***

Ramanujan's approach is now incorporated in computer algorithms, yielding millions of digits of *pi*, when he knew nothing of computer programming.

The turning point

During 1906-1912, Ramanujan was constantly in search of a benefactor, to earn his livelihood. With his 'Notebooks' as his only recommendation, he sought the patronage of V. Ramaswamy Iyer, the founder of the Indian Mathematical Society, who was at Tirukovillur and asked for a clerical job in his office. The former had no mind to smother Ramanujan's genius and sent him back to Madras with a letter of introduction to P.V. Seshu Aiyar, then at the Presidency College, Madras. He gave in turn, Ramanujan a letter of recommendation to that true lover of mathematics, R. Ramachandra Rao, then District Collector, Nellore. This was the turning point in his life.

Ramachandra Rao took a few days to study the notebooks. The latter enjoyed being led "step-by-step to elliptic integrals and hypergeometric series and at last to his theory of divergent series, not yet announced to the world".

To enable Ramanujan continue his mathematical work, he gave regular financial support for an year then Ramanujan joined on March 1, 1912 as a clerk in the Madras Port Trust, with the help of the manager S. Narayana Iyer, himself a good mathematician. He held this post for 14 months; during this period his wife joined him.

Ramanujan's work attracted the attention of prominent Englishmen at Madras — Francis Spring (Chairman, Madras Port Trust), Gilbert Walker (Director General of Observatories), Little Hails and three well-known mathematicians at Cambridge.

Historic letter

On the advice of P.V. Seshu Aiyar, Ramanujan communicated his theorems on divergent series in a historic letter dated January 16, 1913 to G.H. Hardy, who was ten years senior to Ramanujan. With the personal interest of Gilbert Walker and support given by the Indian stalwarts mentioned earlier, the University of Madras awarded its first research scholarship to Ramanujan to study in Cambridge. In quick succession, Ramanujan received in the next three months four long letters from Hardy, who had already sprung into action, advising the India Office of his wish to bring him to Cambridge.

In Cambridge University

Ramanujan left Madras by S.S. Nevasa on March 17, 1914 and arrived in Cambridge on April 18, 1914, Hardy arranged to agument his £250 a year scholarship with the award of an exhibition scholarship of £60 a year.

J.R. Newman, the historian, wrote (1968) "Ramanujan arrived in England abreast and often ahead of contemporary mathematical knowledge. Thus, in ***a lone mighty sweep, he had succeeded in recreating in his field, through his own unaided powers, a rich half-century of European mathematics.*** One may doubt whether so prodigious a feat had ever been accomplished in the history of thought".

Being a strict vegetarian, he had difficulty in getting proper nourishment. Neville, who witnessed his adjustment up close, wrote "He felt the petty miseries of life in a strange civilisation, the vegetables that were unpalatable.... But he was a happy man, revelling in the mathematical society and idolised by the Indian students". In the old town of Cambridge, Ramanujan had found a kind of 'Intellectual Nirvana' (*The man who knew infinity*, Robert Kanigel, Little, Brown & Co. Ltd., London, 1991).

Immediate answer

Ramanujan attended a few lectures. One morning, Mr. A. Berry stood at the blackboard working out some formulae and at one point saw Ramanujan's face glowed with excitement, Berry interrupted to ask him whether he was following the lecture. At this Ramanujan went to the blackboard and, much to everyone's surprise, wrote down some of the results, which were yet to be proven. W.N. Bailey, who attended with Ramanujan a course by Hardy, recalled "I remember that, when Hardy wanted the value of an integral he turned to my colleague, and asked him for the value; and Ramanujan immediately supplied him with the answer... He certainly had a great, effect on my own work: ***if Ramanujan had not lived, I should not now be the new Professor of Mathematics".***

Productive years (1914-1917)

The first three years of his stay, April 1914 to April 1917, despite his strict vegetarian diet, was most productive. He published 21 research papers containing theorems on definite integrals, modular equations, Riemann's Zeta function, infinite series, number theory, modular functions, partitions and combinatorial analysis. His paper 'Highly composite numbers' containing 269 equations was published in the *Journal of the London Mathematical Society*, 1915. For this work he was awarded in March 1916 the B.A. Degree: the domiciliary requirement of six terms was relaxed by the Trinity college. He was elected a college Fellow in 1917, and FRS in 1918. From the spring of 1917, when the first signs of illness appeared, he was in and out of nursing homes until his departure home in March 1919. Ramanujan arrived in Madras on April 2, emaciated and very sick. His wife nursed him till his ***untimely death*** on April 26, 1920.

To mark his 75th birthday, a commemorative stamp was brought out by the Government of India. ***The only Ramanujan museum in the country, founded by Mr. P.K. Srinivasan, a mathematics teacher, operates from March 1993*** in the Avvai Academy, Royapuram, Madras.

His notebooks

Ramanujan's life-long habit had been to work on a slate and transfer the final results to the notebooks. Between 1903 and 1914 he had already three notebooks, which he took to Cambridge. Hardy recalled that Ramanujan was ***showing him half a dozen new results every day*** during 1914-19. He started with one book and left it to G.N. Watson to take up the job of editing the notebooks with B.M. Wilson; but, the project did not materialise for over six decades after the death of Ramanujan (1920).

In 1957, at the instance of Homi Bhabha and K. Chandrasekaran, the ***Tata Institute of Fundamental Research Published a facsimile edition of the notebooks in two volumes.*** These notebooks have formed the basis for numerous papers by many mathematicians all over the world. The stupendous efforts of Bruce C. Berndt who spent over two decades to study each and everyone of the 3,254 entries in the notebooks resulted in the monumental five-volume publication during 1985-1997.

A memorial to Ramanujan

The Srinivasa Ramanujan Centre (SRC) at Kumbakonam is a part of the Shanmuga College of Arts, Science, Technology (SASTRA) a Deemed University. It was inaugurated in 2003 by the President of India, Dr. A.P.J. Abdul Kalam.

SRC is an imposing six-storied structure with a central concrete sloping spiral staircase. A large hall in SRC, named as the Ramanujan House of Mathematics by Dr. Kalam, houses exhibits related to the life and work of Ramanujan, in artistically designed show cases. Two chairs called the DST–Ramanujan Professorships for Number Theory and for Discrete Mathematics were instituted by the Department of Science and Technology (DST), Government of India in 2005.

The academic programme of the SRC is to encourage research activities in the areas of number theory, and the renowned notebooks of Srinivasa Ramanujan in his distinct handwriting.

[*Ramanujan,* K. Srinivasa Rao, East West Books (Madras) Ltd.]

S. Chandrasekhar (1910-1995)

Pushed the limits of astrophysics

He literally went out of the way, driving hundreds of miles from the Yerkes observatory to the University of Chicago to teach a class of only two students. The Noble Prize in Physics for 1957 went to the whole class, Lee and Young.

Educated at Madras and Cambridge

Born on October 19, 1910 in a middle-class family hailing from Tamil Nadu, Chandrasekhar (known as Chandra) had his early education at the Hindu High School, Madras; and in 1930 he obtained his B.A. (Hons) degree in physics from the Presidency College. He made up his mind while in his teens, to pursue ***science as his career.***

Chandrasekhar made full use of the freedom his college allowed, and attended, besides his physics classes, lectures in mathematics. He also studied, on his own, higher mathematics. Like his illustrious uncle, C.V. Raman (1888-1970), he published while still an undergraduate, an original paper in the *Proceedings of the Royal Society,* London.

To know something purely for its own sake and in a manner totally pleasing to himself was his ideal. He pursued the practice of independent studies throughout his life.

Chandrasekhar went to Trinity College, Cambridge, where a few years earlier Ramanujan (1887-1920) set up a glorious record and whose memory he helped to foster. For six years, from 1930 to 1936, he worked on the theory of stellar structure, which laid the foundation for the course of his future work.

He returned home in 1936 and married Lalitha, his collegemate. Cambridge had not given him a permanent position and India had nothing to offer. So in 1937, ***he moved to the Chicago University,*** for what was to prove ***a permanent migration.***

Methodical worker

Chandrasekhar followed a distinctive style in his research work: "He enters into a known area, outlines the present state, tries to add to it and rounds off with a monograph on the subject of study. Then he moves on to another field and masters it", observed Wahli.

His studies were exhaustive and made them appear like there was very little for anyone elso to contribute. But, far from it, ***new directions previously unseen and undreamt of emerged from his works***. Thus, with each monograph representing roughly the culmination of a particular period, Chandrasekhar actually triggered a new field of research.

Chandrasekhar grouped his wide range of investigations under seven periods of his life:

Stellar structure (1929-39); Stellar dynamics (1938-13); Radiative transfer (1943-50); Hydrodynamic and hydromagnetic stability (1952-61); Ellipsoidal figures of equilibrium (1961- 68); Relativistic astrophysics (1962-71); and Theory of black holes (1974-83).

The monographs issued at the end of each of these periods are ***classics in astrophysics.*** For example, the first monograph, published in 1939, is still both a cornerstone of undergraduate courses and an oft-quoted reference in the most abstruse research papers.

Delayed recognition

Chandrasekhar discovered that all stars do not become white dwarfs, by which name densely packed stars in their lives' end are known. Combining Einstein's special theory of relativity and quantum mechanics, he showed mathematically that, if the mass of a star exceeds a certain limit, the star collapses under the gravitational force and creates a 'black hole'. This limit was shown to be a little less than one and a half-times the mass of our sun, and is known as the 'Chandrasekhar limit'.

He announced this discovery in January 1935 at the meeting of the Royal Astronomical Society, London. ***Arthur Eddington*** (1882-1944), who was at that time a leader in the field, ***ridiculed the idea before the scientific community*** and made it look as though Chandrasekhar understood neither relativity nor quantum mechanics. Chandrasekhar was a newcomer in the research area and found himself pitted against an older scientist who was known internationally.

Twenty years later, the Chandrasekhar limit became an established fact in 1955; and it took another 28 years before he recived ***the Nobel Prize for his discovery (1983).***

Formidable as a teacher

Amidst such a background of achievements, almost unrivalled in applied mathematics and astrophysics, Chandrasekhar excelled also ***as a teacher who kept his class spellbound.*** One of his graduate students recalls an incident. His wife, a student of zoology, was sitting in the room next to the hall where Chandrasekhar was teaching. She did not like the physics course she was taking, but she was totally entranced by the rhythm and structure of the sound that came through the wall.

Chandrasekhar was, however, quite formidable as a teacher; frivolous questions from students, who did not appear to have studied the material were dealt with severely. And yet his devotion to his profession was unquestionable. ***There is the celebrated instance when he literally went out of the way to teach a class of only two students.*** In the mid-1940s, he used to drive some hundred miles from the Yerkes observatory to the University of Chicago, week after week, to teach these students. And Chandrasekhar created a record: ***the Nobel Prize in Physics for 1957 went to the whole class, Lee and Yang.***

His adolescent dream

Chandrasekhar is the most platonic example of the ideal scholar, no vanity or even recognition seeking, thorough in whatever he did. Usually one associates long, painful calculations with younger people, but here students had seen him do laborious calculations all by himself, though he was past seventy. He would carry on the work until it was complete aesthetically and not just calculation-wise. Even his rough manuscripts are so well organised and neatly written out, that they could be sent to the press straightaway. He is the only scientist in the world who used Gothic symbols in his equations and one of the few who still used a fountain pen.

Chandrasekhar wrote in a letter to his father, a year after he reached Cambridge, for higher studies. "He is wise who strikes an exact balance between learning for himself and attempting some research consistent with success." With the extraordinary successes he achieved in astrophysics and applied mathematics, Chandrasekhar more than fulfilled his adolescent dream.

Honours

In a career that spanned 55 years, he had become a ***living legend in the field of astrophysics.*** He is the author of several books and the winner of all the international honours — the Royal Medal, Copley Medal and Fellowship of the Royal Society, ***National Medal of Science (U.S.) and Padma Vibhushan (India).*** He worked for nearly 20 years as the editor of the *Astrophysical Journal,* U.S. and elevated it to its present pre-eminence in the professional world. He read and checked every paper published.

(*CHANDRA: A Biography of S. Chandrasekhar,* K.C. Wahli, Penguin India, 1991)

Paul Erdos (1913-1996)

A homeless mathematical luminary

> "For most of his life, Erdos never had a job, but relied on the hospitality of institutions and individuals. Sometimes a leading mathematician would hear a knock at the door at an unearthly hour and find Erdos on his doorstep two battered holdalls in hand, proclaiming 'My brain is open' — a statement meaning he wanted to collaborate for a while..." (*'My brain is open: The Mathematical Journeys of Paul Erdos,* Bruce Schechter, Oxford University Press, 1997)

Turbulent childhood

Paul Erdos (pronouned 'air-dish' was born on March 26, 1913 in Budapest, Hungary. His childhood was a nightmare. As he arrived in the world, being delivered in a hospital, his elder sisters — three and five years old, died of scarlet fever. His parents kept him away from public schools to avoid his contracting contagious diseases. His father and mother, both teachers of mathematics in his schools, instructed Paul themselves, besides hiring tutors.

Paul's father was imprisoned in Russia during World War I and he returned home only in 1920, which was the first time he could talk to his son. ***He introduced Paul to the realm of prime numbers, the building blocks of arithmetic, and also taught him English.***

When Paul was four, he could multiply four-digit numbers mentally. From the age of 13, he participated in the many mathematics contests conducted by *Komal,* a well-known mathematic journal in Hungary for high school students. At the end of every year, his photo appeared amongst the successful candidates.

His contributions

At 17, he found a proof for the celebrated ***Chebyshev's theorem "between any number greater than one and its double, there is a prime number".*** Though he was not the first, to enunciate a proof, this marked his first important discovery. Paul proved some far-reaching generalisations of the theorem, which formed his doctoral thesis. In 1934, he defended his thesis when he was only 2l.

Paul grew up in a strongly anti-communist environment. When he saw the Nazi appetite for territories growing. Paul left Hungary bidding good bye to his parents. During 1934-1938, Paul worked in Manchester.

Migration to the U.S.

Paul was never blind to the crimes commited by the Communist regime. As soon as the first victim of these trials, Laszlo Rajk, was sentenced to death, he packed his suitcases again and fled Hungary; Paul was back in the U.S., but with no assured income, though he was already a famous mathematician.

In 1954, he got an invitation for the International Congress of Mathematicians in Amsterdam. He needed a visa to re-enter the U.S. But, his application was turned down. He could not return for nine years, for his case was reconsidered only in 1963, two years after President Kennedy assumed office.

From August 1955 onwards, Paul spent two months a year in his motherland, but in 1973, for political reasons, these visits had to be abandoned. He returned only in 1976 to visit his good friend and colleague, Paul Turan who was dying.

A vagabond life style

It is not an accident that Paul developed a vagabond life style. He had seen the gruesome restriction, by which his mother was prevented from teaching in a high school for the rest of her life.

He chose to be free, for which he had to make enormous sacrifices. He did not let anyone restrict his free movement in pursuit of his mathematical work, be it Hitler or Stalin or Uncle Sam (the U.S). ***Paul owned almost nothing, not even books.*** He travelled around the world with two battered suitcases containing his minimum worldly possessions.

Paul did not care to own a house; he let his apartment in Budapest be used by visiting mathematicians, old colleagues or friends facing housing problems. He never held a regular job, but rushed immediately to offer help, when any of his friends was in serious trouble. So it was ***his habit to ask people innocently, "How do you make a living?"***

He was ready on a few hours' notice to fly to any place: mathematicians would greet 'Uncle Paul' with delight. Many stored problems to share with him, in the hope that his stay with them would vindicate his motto "***Another roof, another proof";*** and proofs did pour in, spanning areas from probability to number theory and graph theory.

Paul kept producing the most beautiful results, namely his path-breaking papers on probabilistic number theory and on approximation theory. Yet, the Princeton Institute did not renew his fellowship in 1939. John von Neumann (1903-57), the other leading Hungarian mathematician, was a silent spectator.

His political experience had taught him to raise his voice every time he saw injustice was being done either in an individual's private career or freedom. He renounced the honorary doctorate from the University of Waterloo, to register his protest against the unfair dismissal of a colleague.

A missionary

Paul gave a proof (1948) for the 'Prime Number Theorem' which had been studied extensively by Legendre (1752-1833) and Gauss (1777-1855). A friend drew his attention to the work of Ramanujan (1887-1920) who had already finalised an elegant proof of this theorem. He has been hailed as the Western Ramanujan.

Since 1977, Paul Erdos became the best known mathematician in the world. He was in contact with almost all the leading mathematicians of his time. He published joint papers with 458 researchers, but the number of those whom he influenced profoundly is over 1000. He dazzled his acquaintances by his passion for numbers: ***when someone asked about any of his theorems, he could cite the year of publication and the exact page numbers.***

"Wherever he went, his route was marked by a trail of mathematical ideas and discoveries till the end of his life in September 1996. Like a missionary, Paul Erdos found followers everywhere." (*The Mathematical Intelligencer,* Springer-Verlag, New York, 1997)

Stephen Hawking (b.1942)

Who works amidst formidable disability

Today scientists describe the universe in terms of two basic theories — the general theory of relativity proposed by Einstein and quantum mechanics developed by Dirac, Heisenberg and Schrodinger. These two theories stood in splendid isolation.

The general theory of relativity, describes the force of gravity and the large-scale structure of the universe, on scales from a few miles to 10^{24} miles. Quantum mechanics, on the other hand, deals with sub-atomic particles of matter, on extremely small scales, such as a billionth of an inch. One of the major endeavours in physics today, is the search for a new theory incorporating relativity and quantum theory in one set of equations. The quest for the Theory of Everything (TOE) goes on.

At the centre of this challenge lies the work of Stephen Hawking, who is confined to a wheel chair permanently and can communicate only by means of a sophisticated computer-controlled speech synthesiser attached to his wheelchair.

Brilliant student

Hawking was born in Oxford, on January 8, 1942, which is ***exactly to the day, the 300th death anniversary of Galileo,*** the greatest astronomer of the modern era. He obviously holds Galileo as his model, for Hawking says, "Galileo was the first scientist to start using his eyes, both figuratively and physically. In a sense, he was responsible for the age of science we now enjoy".

Hawking entered St. Albans in September 1952, one of England's lesser-known private schools. By the third year, he had come to be reckoned as a bright diligent student of the class. He showed an instinctive insight in solving complicated problems. He would spend very little time on his homework but still obtain full marks. The image of him at this time is that of the schoolboy swot in his grey school uniform and cap as caricatured in *Tom Brown's schooldays.* In the spring of 1958, Hawking and his friends built a computer LUCE (The Logical Uniselector Computing Engine), for which they made use of recycled parts from an old office telephone exchange. This was at a time when only a few universities had computers.

Change in plan of study

His father, a specialist in treating tropical diseases, wanted Stephen to follow him into a career in medicine. All the plans of the former changed, when Hawking won a scholarship to study Natural Science at Oxford. He found the course easy, requiring an average of an hour's work a day. He did not bother about tutorial problems, as ***he could point out errors in the textbooks.*** He fared better at Oxford than at school and ended up in the finals with a first class honours degree.

With an ambition to work for his Ph.D. degree under the legendary Fred Hoyle, who was developing the steady state theory of the universe, he arrived in Cambridge in October 1962. But he ended up working under a relatively unknown professor.

Crisis in health

A personal crisis threatened his course. In early 1963, the first symptoms of ALS, commonly known as motor neuron disease in Britain, appeared. The disease crippled him: doctors gave him little hope beyond two years to live. Hawking fell into a deep depression. ***He owed his recovery to Jane Wilde, his future wife;*** he threw himself to work on his thesis, as he needed a job and, in turn, a Ph.D. for getting a job.

His supervisor took him to London for a discussion with Roger Penrose. In the train on the way back to Cambridge, ***he got a flash inspiration to apply Penrose's singularity theory to the entire universe.*** This formed the brilliant, frail chapter of his thesis, which earned the ailing 23-year old Hawking his Ph.D. degree. In July 1965, he was married to Jane in the Chapel of Trinity Hall. They have three children. As one of his close friends observed, while ***introducing Hawking to a Chicago audience: "not all of Stephen is paralysed!"***

Breakthrough in cosmology

Hawking began spending more time on Hoyle's theory than on his own work. When Hoyle made public his findings at a meeting of the Royal Society, London, Hawking challenged one of Hoyle's calculations and answered defiantly that he had worked it out, with different results. Thus Hawking began to make a name for himself.

In 1974, Hawking made the revolutionary proposal of exploding black holes, made a bold attempt to link quantum mechanics and the general theory of relativity. The radiation emitted by blackholes was called 'Hawking radiation' (*Nature*, March 1, 1974).

Hawking found that a black hole cannot only shrink, it can also throw off energetic particles generated in its intense gravitational field. After a long enough period of time, say, a few thousand million years, the hole will get hotter and hotter, until finally it bursts, ***the singularity inside the hole exploding in a fireball of high-energy radiation called a Big Bang.*** Hawking was elected a Fellow of the Royal Society in 1974.

Reviving popular interest

His book A *Brief History of Time* has sold over ten million copies worldwide since its publication in 1988. This was filmed in January 1990. By this book, he revived popular interest in scientific work and showed his prodigious talent for expressing complex theories in simple words. ***His pithy description of black holes is "they can eat the whole solar system for breakfast!"***

Hawking is now ***the Lucasian Professor of Mathematics at the University of Cambridge,*** a post, first held by Newton and later by Dirac, two celebrated explorers of the very large and very small. ***Hawking has travelled around the world, giving talks and showing legendary skill in explaining.***

Hawking has paid a heavy price for his success. The seemingly unthinkable about their marriage happened in spring 1990. A few months short of the silver jubilee of their wedding, the Hawkings separated. Religious incompatibility is said to be one of the factors.

Einstein's heir

Hawking has made fundamental breakthroughs in cosmology and has pushed forward our understanding of the universe we live in. He ***has continued to work*** in spite of a crippling disease, which has made him ***paralysed and speechless.*** For his work on black holes and the big bang theory, he has been acclaimed Einstein's heir.

Honours

From 1977, his achievements began to attract wide publicity. Honours and prizes were heaped on him. He received the Einstein award in 1978 at a gala event in Washington. He has been made a CBE and then Companion of Honour (1989). He has been awarded honorary doctorates by five universities. Cambridge and Leicester in the U.K. and New York, Notre Dame and Princeton in the U.S. He received the Medal of Freedom (U.S.), 2009.

Mendeleev

SECTION FOUR

Liebig

CHEMISTRY

Hodgkin

Joseph Black (1728-1799)

Founder of calorimetry

Temperature had been measured by thermometers since the time of Galileo (1564-1643): each investigator was able to mark on his home-made thermometer, points corresponding to definite temperatures. The temperatures of melting ice and boiling point of water were brought into use as fixed points.

But a different question remained, that of stating the quantity of heat, which is needed to raise the temperature of a given body by a given number of degrees. This question appeared at that time insolvable; it was Joseph Black who created the necessary foundation for dealing with all questions of heat quantity by an extensive series of observations based upon fundamental ideas.

From medicine to chemistry

Joseph Black was born on April 16, 1728 in Bordeaux, France; but by blood, he was a pure Scot. He, the fourth in a family of 12 children, was first educated by his mother before being sent to a private school at Belfast. Like so many Ulster Scots, he migrated in 1744 to the University of Glasgow. He elected to study medicine, on the advice of his father to choose a profession. Though he began the study of anatomy, ***lectures in chemistry by William Cullen produced a decisive influence on shaping his future work.*** For three years, he served as Cullen's assistant, which cemented a close friendship.

Flourishing medical practice

In 1752, Black shifted to the more prestigious University of Edinburgh, which had on its faculty of medicine eminent teachers — the great anatomist Alexander Monro, Robert Whytt, a physiologist and Charles Alston, a chemist. He gained more from the bedside clinical instruction than from the lectures. ***Black was a discerning student***; though the lectures of Charles Alston pleased him, he could find them deficient in chemical knowledge. ***For he wrote to Cullen, "no branch should be more cultivated in a medical college".***

In 1754, Black received the M.D. degree for his classic dissertation 'Experiments upon magnesia alba, quick lime and some other alkaline substances'. This was published in 1756 by the Philosophical Society of Edinburgh. This was his only major publication; it established his scientific renown. The same year he succeeded Cullen as professor in Glasgow at the age of 28. Here, Black spent the next 10 years, when he developed ***his ideas about latent and specific heats.*** He carried out experiments, alone or with his students, to confirm his theories but he did not publish his discoveries. Alongside his teaching courses in chemistry, he carried on a flourishing medical practice.

In 1766, Black succeeded his own professor again at Edinburgh, where he was destined to spend the rest of his career. He limited his medical practice to the care of a few close friends like David Hume. ***His period of scientific creativity came to an end.***

Shift to teaching

The teaching of chemistry now became his passion: here, as in Glasgow, he became an idol to medical students. Students flocked from America, Germany, Russia, Switzerland and Scandinavia. The illustrious among them were, Thomas Hope who succeeded him in 1797; Daniel Rutherford, discoverer of nitrogen; John McLean, who became in 1795 Princeton's first professor in chemistry; Thomas Young, physician–physicist; Thomas Thomson, pioneer historian of chemistry.

Teaching style

Black emerged as ***a gifted and effective teacher.*** He spoke almost extempore, from the barest notes; his tone was conversational, without any show of flamboyance. ***His lectures,*** of which numerous manuscripts by his students have survived, ***were models of precision and clarity.*** John Robinson, his student at Glasgow, brought out in 1803 his master's *Lectures on Chemistry.*

Black's investigation of alkaline substances had a medical origin. At Edinburgh, he found it expedient to avoid any conflict between two of his professors, Robert Whytt and Charles Alston, who were investigating lime water for its presumed efficacy in dissolving stones (urinary calculi).

He chose instead 'magnesia alba', a white powder which was known as a mild purgative. He showed by careful quantitative experiments that this substance, a mild alkali, lost weight in heating and this loss in weight was due to the release of 'unfixed air', now known as carbon dioxide. ***Black's application by the chemical balance to the study of such chemical reactions*** was one of his valuable contributions to science.

Fundamental discoveries

In the experiments on heat that Black performed in the summer of 1761, he found that the change of state, for example melting of ice or vaporisation of water, of a given mass of a substance does not only require a certain temperature, but also a certain quantity of heat, which is always consumed, thought it cannot be detected by the thermometer.

Black was not satisfied with mere qualitative demonstration but measured the heat of fusion of ice and the heat of evaporation of water. He called them the 'latent heats' and ***his results are as near to the figures known to us today.*** He described them regularly in his university lectures, whence the ideas spread.

Black is known for the fundamental notion of the ice calorimeter, which was later refined by Robert Bunsen (1811-1899) and became one of the exact methods of measuring heat.

The second fundamental discovery made by Black was 'specific heat'; that different substances have different heat capacities. He showed that mercury, despite its greater density, must have a smaller store of heat than an equal amount of water at the same temperature.

When Black started his career in Glasgow***, James Watt (1736-1819) was taken under his wing as a technician.*** The latter was astonished at the large amount of cold water required to condense the steam engine in the cylinder, ***until Black explained his ideas about latent heat of steam***. This clarified the problems Watt encountered, which in turn, aroused Black's interest.

Chronic illness

Black never married, but he was no recluse. He enjoyed conviviality: he never declined company to his poor patients or his students at his table. He suffered all his life from chronic ill health, perhaps pulmonary in origin. He nevertheless managed by careful diet and moderate exercise to husband his strength.

Black gave his last full course of lectures, at the age of 67 in 1795-96; but aware of his debility, he choose Thomas Hope as his successor. He lingered on for two more years. The manner of his death (December 6, 1799) was so peaceful that it has been several times recounted.

Joseph Black's view of chemistry, his chemical doctrines, can be derived from his single major paper and from his lectures, which he left behind in full manuscript form. To him, ***chemistry was a subject with wide practical application to medicine and to the progress of industry.***

(*Dictionary of scientific biography,* Vol. II, Ch. Scribner's Sons, New York)

Antoine Fourcroy (1755-1809)

From medicine to applications of chemistry

His most important contribution to the progress of chemistry was the advocacy of the theories of Antoine Lavoisier (1743-1794).

Fortunate turn

Antoine Fourcroy was born on June 15, 1755 in Paris, the son of an apothecary. He did not complete his studies at college but took up a ***job the age of 16, as a copying clerk in the office of the Government Chancellery.*** There, he would have spent his life but for the good fortune of meeting a well-wisher, who persuaded his father to let the young man study at the Faculty of Medicine.

Fourcroy graduated as a doctor in 1780 but did not practise medicine. He gave lectures in chemistry in the private laboratory of his teacher Bcquet and after the latter's death, in his own laboratory during 1780-1792. ***In his lectures, he emphasised the relation between chemistry and natural history and their application to medicine.*** He decided to concentrate on chemistry.

Advocacy of Lavoisier's views

At first he was as uncertain as many of his colleagues, but after the establishment of the composition of water and nitric acid by Henry Cavendish (1731-1810), he gave it his full support; Fourcroy wrote many works, the one based on his lectures running to many editions in several languages.

He classified the chief facts of chemistry under twelve headings, which he published as a little book *Philosophie Chemique* (1792); this proved to be popular and was translated into 11 languages.

Fourcroy's interest in the applications of chemistry to medicine led him to study various solids and fluids of human and animal body in health and sickness.

Like most French scientists, Fourcroy supported the moves that led to the French Revolution. He entered into politics in 1789, then in 1792. He served on the committee that founded two famous institutions, Ecole Polytechnique and Ecole de Medicine at Paris. He served as professor at each institute till his death (December 16, 1809).

Application of his chemical expertise

The French Government called on scientists to assist in solving the country's economic difficulties. Fourcroy responded to the call in the following practical way.

In 1790, Fourcroy successfully applied his chemical expertise to the problem of extracting copper (needed for coinage and ship building) from its alloy tin, which, with the closing of many churches, was available from the church bells. He heated the molten metal in air, until the gain in weight showed that enough oxygen had been absorbed to oxidise the tin but not the copper, he continued the heating next in the absence of air.

By this technique, all the oxygen was transferred to tin. The process gave a good yield of copper and was employed on a large scale for at least a decade.

Fourcroy was responsible for organising a commission of scientists that made it possible to develop the metric system of weights and measures.

Contribution to inorganic and animal chemistry

Fourcroy and Vacquelin gave a clear description of the preparation and properties of sulphites and phosphates (1796). With Thenard, he undertook a masterly study of the oxides of mercury, which established the existence of two series of compounds containing mercury in different degrees of oxidation.

Fourcroy examined many solids and fluids, including brains, mucus, bile and so on, and tried to explain their formation and function in chemical terms; and to find remedies that would restore them to their original state.

Fourcroy expected that the analysis of urinary calculi would lead to the discovery of solvents suitable for dissolving them by injection into the bladder, but this was not achieved.

In an attempt to find why urinary calculi were formed, Forucroy and Vacquelin investigated urine. They gave ***the first satisfactory account of 'urea', which name they coined in 1799.***

Honours

As a councilor of state, Forucroy played a large part in drafting a new educational system, from schools to advanced colleges. Napolean appointed him director general of public instruction (1802) and conferred on him the title of Count of the Empire (1808).

Antoine Lavoisier (1743-1794)

Founder of modern chemistry

Though Lavoisier adopted the slogan of the revolution, it was an irony of fate that his head was claimed by the guillotine.

Early education

Antoine Lavoisier was born in Paris on August 26, 1743, the son of a prosperous lawyer. He began schooling at the age of 11 at the College Mazarin in Paris, where his scientific interests were aroused. Besides, he graduated in law in 1763 to follow his father's profession, but turned to geology on the advice of a family friend. He helped the latter to map the geology of France.

As a tax collector

In 1768, Lavoisier received a provisional appointment to the French Academy of Sciences. At the same time, he invested in a private tax collecting company, which enabled him to set up a research laboratory for conducting his experiments. But this association as a tax gatherer proved to be a fateful step. He rose in the Academy of Sciences, becoming a director in 1785 and treasurer in 1791.

Public offices

Lavoisier's public duties covered the role of the barrister, tax official, financier, agricultural scientist and director of the gunpowder administration. In between these activities, he devoted ***one day a week exclusively to the pursuit of science.*** He was supported in his scientific career by his wife Marie Paulze whom he married in 1771, when she was not quite 14. She translated English chemical works into French and assisted him in the laboratory. At the weekly scientific gatherings, she was an indefatigable promoter of the new chemistry.

In Lavoisier's time, development of chemistry was thwarted by a belief in phlogiston theory, that combustible substances contain 'phlogiston' which they lose when they burn. His research between 1772 and 1778 established that oxygen was an active constituent of air and was absorbed by metals and non-metals on burning.

His research was largely devoted to developing a new theory of combustion, which ***transformed the language of chemistry.*** He was helped by the Academy of Sciences, which loaned a large burning lens. "His delight in having access to the lens was exactly like that of the twentieth century physicists who run experiments on gigantic particle accelerators: it was a machine capable of blasting apart substances previously considered immutable" (*Antoine Lavoisier — Science, Administration and Revolution,* Arthur Denovan, Blackwell Series, 1993).

New system of chemical nomenclature

Lavoisier showed that water was not an element, as believed for over 2,000 years, ***but a compound of hydrogen and oxygen***. He synthesised water by burning jets of hydrogen and oxygen in a bell jar over mercury, which was an important landmark to his combustion theory. In 1789, Lavoisier published his work *The Elementary Treatise on Chemistry,* the core of which was the oxygen theory. The book became an effective vehicle for the transmission of his new system of chemical nomenclature. For example, acids regarded as compounds of various elements with oxygen were given names, which indicated the element together with its degree of oxygenation. It has been translated into several languages and it remains a classic in the history of science.

Lavoisier extended ***his theory of combustion to the area of respiration physiology.*** The result was published (1782) in a famous memoir 'On Heat', in collaboration with Laplace, the celebrated mathematician. He designed a set of experiments to study the whole process of body metabolism and respiration. Lavoisier's pioneering work in this field served as a pacesetter for research work on physiological processes for generations to come. In 1775, Lavoisier was appointed director of the Royal Gunpowder Administration. He introduced successful reforms, which increased the Government revenue. He was made, in 1790, secretary of the commission to secure uniformity of weights and measures. This led to ***the establishment of the metric system throughout France***.

Tragic end

As the French Revolution gathered momentum from 1789 onward, attacks mounted on tax-gatherers. His enemies accused him of plundering people and the state treasury. Though Lavoisier adopted the slogan of the revolution, 'everything must be destroyed, since all must be created afresh', it was an irony of fate that his head was claimed by the guillotine on May 8, 1794. A close friend of Lavoisier, the great mathematician Joseph Lagrange cried:

"Only a moment to cut off that head and a 100 years may not produce another like it".

Claude Berthollet (1748-1822)

Who founded applied chemistry

A more successful innovation for putting chlorine to practical use was his introducing it as a bleaching agent for textiles. Till then, the method of bleaching involved soaking cloth in whey and spreading it in a sunny field. This prevented large tracts of land from being tilled and was wasteful. Berthollet demonstrated in public the application of his method.

Early years

Berthollet was born on December 9, 1748 (in a village of Savoy district, France) at a time, when the family was in straitened circumstances. He attended the University of Turin in Italy and qualified as a physician in 1768. Then he presumably practiced medicine for four years till 1772, when he moved to Paris. As his Italian degree was not recognised, he continued to study medicine and took in 1778 a second degree in medicine from the University of Paris. ***Besides medicine, he studied chemistry.***

By 1780, his research work in chemistry earned him membership to the French Academy of Sciences. In 1785, Berthollet adopted a new system of chemistry based upon the oxidation theory of combustion developed by Lavoisier (1743-1794) who was his mentor.

Berthollet ***flourished under four different political regimes***. He served in 1792, as a member of the commission for the reform of the monetary system and the commission for war production. He was one of the founders of the famous Ecole Polytechnique at Paris.

Career under Napoleon

Berthollet's career reached its zenith under Napoleon who showed genuine admiration and affection for him. The latter appointed him to accompany the commission that was to bring back the great works of Italian art to France.

During this assignment, he showed his talent for originality by developing some of the earliest chemical methods for the restoration of paintings. He accompanied (1798) Napoleon on the expedition to Egypt, where he and Monge (1746-1818), his lifelong friend, stayed for two years. Napoleon heaped honours on Berthollet — a count, the administrator of the mint and officer of the Legion of Honour.

After 1804, he retired to his country home at Aruceil, near Paris. This became the centre for discussing scientific problems, which blossomed into ***the famous Society of Aruceil (1807) under the direction of Berthollet and the great mathematician Laplace (1749-1827).*** His last days were clouded by the suicide of his son in 1810, following the failure of his chemical factory. Berthollet died at Aruceil on November 6, 1822.

Unlike his mentor Lavoisier (1743-94), Berthollet re-invigorated the traditional science by synthesising ideas rather than revolutionising the basis of science. Trained as a physician, he strove to find a theoretical explanation of chemical phenomena. He worked hard to put science at the service of man's practical needs.

Apostle of Lavoisier

Berthollet made many original contributions to theoretical and applied chemistry. He reformulated the basic principles of chemistry by synthesising the traditional views with the important new discoveries of Lavoisier, which took him the better part of twenty years. During this period, he rejected the traditional 'phlogiston theory' and became literally the apostle of Lavoisier, preaching the latter's doctrine.

From this followed ***the determination of the composition of ammonia, for which he gave the earliest accurate analysis (1785)***.

Berthollet investigated methods to obtain, by replacing nitre with potassium chlorate, a more powerful kind of gunpowder. He gave in 1788 a public demonstration with a mixture of pottasium chlorate and carbon, which resulted in the death of five people on the spot. ***This innovation later found its way to military operations.***

Contributions to textile industry

It gave concrete proof of his humanitarian impulse that, unlike his contemporaries such as James Watt who amassed large fortunes from their industrial inventions, he gave gratis his method.

His method of chlorine bleaching involved pouring sulphuric acid on a mixture of six ounces of manganese monoxide and sixteen ounces of salt; this mixture was heated by immersion in boiling water. This was his discovery of bleaching liquid, which he named it 'lye de Javelle'.

Berthollet's another contribution to the textile industry was the treatise in which he tried to place the ancient craft of dyeing on a scientific basis, by a systematic discussion of its procedures and enunciation of an adequate set of theoretical principles to explain the chemical reactions involved.

Fundamental work

Berthollet made original objections, when many new situations arose in the application of Lavoisier's ideas. A noted instance was the latter's explanation of the properties of acids as deriving from the presence of oxygen. He had already shown in 1778 that hydrogen sulphide, while possessing the characteristics of a feeble acid, did not contain oxygen. He reaffirmed this discovery twenty years later. Again by his ***analysis of prussic acid (1787), he correctly showed it to be composed of hydrogen, carbon and nitrogen. This was one of Berthollet's fundamental contributions.***

Theory of affinity

During his stay in Egypt, Berthollet noticed the apparently inexhaustible source of sodium carbonate at the threshold of the desert. He gave an explanation for this natural phenomenon in terms of a chemical theory of affinity, that had been maturing in his mind over the years.

Berthollet attributed it to decomposition of the salt, assuring a constant production of sodium carbonate and calcium chloride through double decomposition, as a result of the redistribution of the affinities between the original reactants.

Berthollet maintained that such physical conditions as the mass temperature, relative concentration and quantities of the substances affected the nature and direction of affinities in a chemical reaction. This led Berthollet to formulate his complete new system of chemistry in his two important works 'Researches into the Laws of Chemical Affinity (1801)' and 'Essay on Chemical Statics'.

(*Dictionary of Scientific Biography,* Vol. II, Ch. Scribner's Sons, New York.)

John Dalton (1766-1844)

Who charted the course of chemistry

It is his atomic theory that gave chemistry a charter as a separate science.

Teaching in schools

John Dalton was born on September 6, 1766, in an isolated village in the north of England. His father was a weaver and a Quaker. He went to school till he was 11. When he was 12, he took charge of a Quaker school in Cumberland. Two years later, he took up teaching in a school in Kendal, where he remained for 12 years. By 1793, he settled down in Manchester. He said, ***"My head is too full of triangles, chemical processes and electrical experiments*** to think of marriage". He worked as a teacher of mathematics and natural philosophy at the New College, established by the Presbyterians.

Meeting his mentor John Gough

Dalton resigned after six years to take charge as Secretary of the Manchester Library and Philosophical Society. Thereafter, he worked as a private tutor of mathematics and chemistry, pursuing it as a career to meet his modest needs. In 1817, he became President of the Philosophical Society, an honorary office that he held for 27 years. Here he met his mentor, ***the blind Quaker philosopher John Gough (1757-1825), whose learning and enthusiasm gave him great stimulus.***

Under Gough's tuition, Dalton made rapid progress in mathematics, meteorology and botany. He began, in 1787, to write a diary keeping meticulously records of the local weather conditions — pressure, temperature, wind velocity and humidity and maintained them for 57 years till his death on July 27, 1844. ***His concern with rain and water content of the atmosphere*** in the wet Lake District seems to be ***the origin of his work on mixture of gases and atomic theory.***

His atomic theory

He was able to explain why the constituents of a gaseous mixture remain homogeneously mixed, instead of forming layers according to their density. In the paper on gas solubility that he presented (1803), he appended rather casually ***the first table on atomic weights***. His public lectures and studies in meteorology were brought out as a book, *A New System of Chemical Philosophy* (1809). In the second part of the book that was published in 1810, he finally launched the chemical atomic theory and showed the practical power of his approach.

The systematic utilisation of Dalton's theory was to be plagued for many years by methodological problems, which were slowly resolved through the work of Gay-Lussac, Avogadro and Cannizzaro. His ideas on real physical existence and actual nature of chemical atoms initiated a continuing nineteenth century debate that came to a close with the work of Rutherford and Soddy.

The celebrated laws

By 1805, Dalton had completed in essence, the outline of his work by which he made his reputation in science: the law of expansion of gases at constant pressure (now known as Charles law), the law of partial pressures which states that the total pressure of a mixture of gases equals the sum of the pressure of the gases, "each gas acting as if it is present alone".

Dalton formulated the ***Laws of Definite and Multiple Proportions,*** which state that a chemical combination of different elements occurs in simple ratios by weight (for example, elements A and B would form a compound AB, possibly A_2B, AB_2 and A_2B_3). ***These laws gave a new basis for all quantitative chemistry.***

Other investigations

Besides the studies in meteorology which he presented as papers before the Manchester Philosophical Society, Dalton investigated other topics: barometer, hygrometer, the formation of clouds, rainfall, evaporation and characterisation of atmospheric moisture including the concept of dew point. He was the first to confirm the ***theory that rain is caused not by any alteration in the atmospheric pressure but by a fall in temperature.***

Dalton was accused of being a very coarse experimenter, as ***he worked with simple apparatus, which he himself constructed.*** But he had a superb gift for reasoning and for drawing correct conclusions from imperfect experiments. It is only his Atomic Theory, that every element consists of very small particles of matter called atoms, which are identical in all respects including mass, that gave chemistry a charter as a separate science.

Honours

Dalton received many honours — FRS (1826), Member of the French Academy of Sciences, honorary doctorates from the Universities of Oxford (1832) and Edinburgh (1834). He was also the co-founder of the British Association for the Advancement of Science. The city of Manchester recognised his services by raising in 1834 his statue.

AMEDEO AVOGADRO (1776-1856)

FROM LAW TO PHYSICS AND CHEMISTRY

He did not support his results with any impressive accumulation of experimental results. It was almost fifty years later that the same result was obtained conclusively from the kinetic theory of gases by Robert Meyer and Rudolf Classius. Avogadro's hypothesis is the basis of all modern chemistry.

Finding his vocation

Amedeo Avogadro was born on August 9, 1776 in Turin, Italy, in a family of well-established lawyers. It is believed that the name itself was probably a corruption of the Latin word 'Advocarii'. He took a degree in philosophy in 1789, another in law in 1792 and a doctorate in ecclesiastical law in 1796. After several years of legal practice, ***he found his true vocation in science*** by studying privately mathematics and physics. The first scientific work he undertook was on electricity in 1803, probably inspired by the then recent discoveries of his contemporary, Volta (1745-1827).

Avogadro started his career in 1806 as a demonstrator in a college and became, in 1820, professor of mathematical physics when the University of Turin was established. Due to the political turmoil that gripped Italy, in 1821, the chair of mathematical physics was suppressed. When it was revived in 1832, he returned to the position in 1834, succeeding the French mathematician Cauchy (1789-1857) and held it till his retirement in 1850.

The hypothesis

At the time when Avogadro published his class memoir (1811), the terms 'atom' and 'molecule' were used loosely. Berzelius (1779-1848) had substituted atom for volume in cases of combining gases and had ended with the logical contradiction of half an atom. Avogadro overcame this difficulty by introducing a polyatomic molecule, but this concept was quite novel. He suggested that besides molecules that contained only one element; there might be molecules that contained different elements. ***This postulate of a compound molecule provided the much-needed key*** to the problems of nineteenth century chemistry, by distinguishing between atoms and molecules.

Avogadro did not support his hypothesis with any impressive accumulation of experimental results. It was almost 50 years later that the same result was obtained conclusively from the kinetic theory of gases by Robert Meyer and Rudolf Classius. ***The statement "equal volumes of all gases, under the same temperature and pressure, contain equal number of molecules"*** is now known as Avogadro's law. It is ***the basis of all modern chemistry.***

By Avogadro's hypothesis, molecular weights — and therefrom atomic weights — can be determined directly, as the relative weights of the molecules of any two gases are the same as the ratios of the densities of these gases, under the same conditions of temperature and pressure. This enables the chemist to deduce atomic weights without recourse to Dalton's arbitrary rule. Avogadro acknowledged that "his hypothesis was at bottom merely Dalton's system, now furnished with a new means of precision from the connection we have found between it and the general theory established by Gay-Lussac" (1778-1850).

Modern formulae

So, Avogadro was able to derive the modern formulae for water vapour, nitric oxide, nitrous oxide, ammonia, carbon monoxide and hydrogen chloride. In 1814, he published the correct formulae for several compounds of carbon and sulphur.

When experimental evidence was lacking, he reasoned by analogy, as in the case of silica, namely SiO_2, in comparison with CO_2. ***He gave molecular***

weights using modern symbols for a host of metals — mercury, iron, silver, lead, aluminum and silicon.

After deducing the formulae of such inorganic compounds as the oxides of nitrogen and phosphorus from combining volumes and densities, or merely by analogy, he turned to organic chemistry, where he gave the formulae for turpentine ($C_{10}H_{16}$), alcohol (C_2H_6O) and ether ($C_4H_{10}O$).

Avogadro stated his hypothesis repeatedly over the period 1811-1821 but ***it was disregarded until 1858*** by a majority of chemists. He had the habit of constantly giving reference only to his own earlier work: but, by establishing the lineage of his research, he unwittingly contributed to his comparative obscurity.

It was Cannizzaro (1826-1910) who made Avogadro's hypothesis widely known at the Karlsruhe Congress (1860), how it would solve many of the major problems of chemistry. In particular, he clarified the relation between atom and molecule, which Avogadro had not made explicit. He determined the atomic weights of many elements. For the first time, there began to exist among chemists substantial agreement on atomic weights.

Memoirs on electricity

Avogadro saw ***no boundary between physics and chemistry,*** as shown by his various studies on heat and his thermal studies of gases. His first two memoirs (1806, 1807) were on electricity: he considered the state of non-conductor placed between two oppositely charged plates, which gave proof that he had some conception of the polarisation of dielectrics. At the end of the memoir, Avogadro suggested that the capacity of a condenser was independent of the gas between the plates and there would be the same process of induction even in a vacuum, a phenomenon later verified by Clerk Maxwell (1831-1879).

Avogadro Number

Named in his honour, it is the number of molecules per gram molecular weight of a substance. The earlier scientists Dalton (1807), Ampere (1814), Dumas (1827) and Prout (1834) independently formulated the same hypothesis. To

Avogadro alone, goes ***the credit of applying this hypothesis to the whole field of chemistry.*** He stands as one of the founders of physical chemistry.

(*Dictionary of Scientific Biography,* Vol. 1, Charles Scribner's Sons, New York.)

Joseph Gay-Lussac (1778-1850)

Scientist who worked with his hands

In addition to his fundamental research work, he contributed to the improvement of industrial chemical processes, notably adding to the lead chamber process, the sulphuric acid tower.

An able boy without fortune

Joseph Gay-Lussac was born in St. Leonard, France, on December, 6, 1778, in a family of five children. His private education received a setback due to the French Revolution, as his father was arrested as a suspect. The opening of the Ecole Polytechnique at Paris however, provided the opportunity for an able boy without fortune.

From engineering to chemistry

He studied engineering and graduated in November 1800, but thereafter became interested in physics and chemistry. The chemist Berthollet (1748-1822), impressed by the capability of the boy, took him to his country house at Aruceil and gave him laboratory training. The former played a key role in the professional development of Gay-Lussac. From 1803 to 1809, he served as adjoint faculty in the Polytechnique. He was professor of physics in the University of Paris till 1832, when he shifted to the National Museum as professor of general chemistry.

Gay-Lussac became well-known through his hot-air balloon ascents in 1804. His objective was to see whether the magnetic intensity at the earth's surface decreased and whether the composition of air changed with an increase in altitude. In 1808, ***he married Josephine,*** then a seventeen year-old shop assistant ***whom he had seen reading a chemistry book between serving customers.*** He sent her to school to complete her education in chemistry.

Fundamental research

Gay-Lussac's first major research was on thermal expansion of gases — he improved the earlier work by taking precautions to exclude water vapour from his apparatus. After repeating each experiment several times with a variety of gases, he concluded that equal volumes of all gases expand equally with the same increase in temperature. This was first observed by Charles in 1787. The law of thermal expansion is known as Charles Law.

The next problem he perused with Alexander Von Humboldt, who was ten years senior to him, was to determine accurately, the relative proportion by volume in which hydrogen and oxygen combine to form water. He clearly expressed his preference for volumes to weight. This is one of his important contributions known for its accuracy as ***he established that the presence of moisture, which would be difficult to estimate by weight, did not alter the volumetric ratio***.

At a meeting of the Society Philomatique in Paris, ***Gay-Lussac announced on December 31, 1808, the Law of Combining Volumes:*** gases combine in the ratios of small integers and when the result of the union is a gas, its volume has also a simple relation to those of its components. This law was a natural consequence of his work with Humboldt (January 1805), when he had found that 100 parts by volume of oxygen combine with almost exactly 200 parts of hydrogen.

Gay-Lussac's interest in volumes led him to devise ***an apparatus by which the vapour densities of liquids could be compared.*** This apparatus was improved 50 years later by Hofmann (1818-92) and the technique is now known as Hofmann's method.

Collaboration with Thenard

Some of Gay-Lussac's best work was done in collaboration with Louis Thenard (1777-1857), the French chemist who laid the foundation of organic analysis. Stemming from Davy's method of electrolysis, which had yielded only tiny amounts of potassium and sodium, they produced 25 g of the alkali metals by purely chemical means, fusing the alkalis with red-hot iron filings. Napoleon made considerable sums of money available to support their work on electrolysis also.

Inspired by the work of Volta (1745-1827), Gay-Lussac and Thenard ***investigated the chemical effects of electric current and showed that the rate of decomposition of the electrolyte depended on only the strength of the current*** — and not on the size of the electrodes. They received the Royal Institution's prize for this work, ***which was 30 years before Faraday.***

With Thenard, Gay-Lussac discovered the amides and oxides of alkali metals and isolated the element boron. In 1809, they made the dangerously reactive gas fluorides — hydrogen fluoride and boron trifluoride. Gay-Lussac was temporarily blinded by a potassium explosion, which demolished his laboratory, but he never lost his enthusiasm for experimentation.

In 1814, Gay-Lussac published his research on the new element iodine (which was discovered by Courtois) and demonstrated the analogy of its properties with those of chlorine. ***He was the first to prepare cyanogen*** and showed it to resemble the halogens and to be the parent of a series of compounds, the cyanides.

Chemical analysis

Gay-Lussac made a pioneering contribution to chemical analysis, in 1832, when he introduced a volumetric method for estimating silver by which he replaced the centuries-old method of cupellation. From then on, ***the concept of titration*** passed into general chemical practice. His work on the properties of sulphates and sulphides was an important step in the perfection of the general system of analysis which later blossomed into 'volumetric analysis'. ***He was the first to use the terms 'pipette' and 'burette'*** that have since become standard pieces of apparatus.

Gay-Lussac's Tower

In addition to inventing laboratory devices — portable barometer, steam injector pump and air thermometer — GayLussac contributed to the improvement of industrial chemical processes, notably adding to ***the lead chamber process the sulphuric acid tower*** which bears his name. He carried out fundamental studies on the effect of light on chemical reactions, on chemical equilibrium and the reliance to a reaction of the mass of the reactants and the relationship between the specific heats of gases and their densities.

Honours

Gay-Lussac's life was entirely devoted to pure and applied chemistry, but he did have a brief political stint. He was elected to the Chamber of Deputies in 1831, 1834 and 1837. He was elevated in 1839 to the rank of a peer; but this honour had been delayed for 17 years, for ***the old aristocracy argued that Gay-Lussac worked with his hands.***

Gay-Lussac died in Paris on May 9, 1850. He remains one of ***chemistry's immortals like Lavoisier, Berzellus, Davy and his own pupil Liebig (1803-73).***

(*Chambers Concise Dictionary of Scientists,* W & R Chambers Ltd., Edinburgh, 1989)

Jacob Berzelius (1779-1848)

From medicine to chemistry

Early in the 19th century, scientists were becoming aware that chemical elements combined in small ratios, which would always be in whole numbers. To this hypothesis, known as Proust's Law, Berzelius supplied the theoretical proof based on the Atomic Theory of Dalton. He devoted his time to analyse a large number of salts and verified this concept. Thus, the Law of Constant Proportions in which the constituents of inorganic compounds are bound together was enunciated.

Study of medicine amidst hardships

Berzelius was born on August 20, 1779 in Linkoping, Sweden, in a family of clergymen. His father and mother died in his boyhood, before he completed ten years. He was sent to the home of his maternal uncle: ***to escape the conflict in the family, he moved in 1794 to a nearby farm and worked as a tutor.*** Two years later, he began his studies at the medical school at Uppasala, which he had to interrupt again to earn money as a tutor. In 1802, he gained the M.D. degree, submitting a thesis on the study of the effect of the galvanic current on patients with different diseases, which work boosted his interest in electrical phenomena.

His uncle, to get rid of Berzelius, arranged for the latter being appointed as an unpaid assistant at the College of Medicine in Stockholm. Here Berzelius lived in the house of Wilhelm Hisinger, a wealthy mine owner. The chemical investigations he began to undertake with this enthusiast in mineralogy and chemistry laid the foundations for Berzelius' future work. ***His meager income was from the job as physician to the poor*** in and around Stockholm. The business he attempted ended in failure and he ran into serious financial difficulties. His financial position improved in 1807, when he became professor of medicine and pharmacy. He now had access to a laboratory and was able to devote most of his time to teaching chemistry.

Changing to chemistry

Berzelius was almost ***self-taught in chemistry.*** Realising the dearth of Swedish textbooks on chemical subjects, he began to work on the preparation of his *Textbook of Chemistry.* He published in 1808 the first volume and, after carrying out further analytical studies, the second volume in 1812. ***The book became the most authoritative chemical textbook of its time*** and was translated into Dutch, French, German, Italian and Spanish. The composition of the book and the problems that occurred to him while writing it strongly influenced the direction of his later career.

Berzelius had to build the scientific apparatus he needed, and to synthesise his own reagents. The new forms of apparatus he constructed were described in his textbook and became standard in laboratories all over the world.

The Law of Constant Proportions

Berzelius also worked out new analytical methods, in the planning of which ***he displayed his genius.*** For example, he determined the ratio of metal to oxygen in a number of metallic oxides by reducing the oxide to the metal or by converting the metal to its oxide. Similarly, he determined the oxygen to sulphur ratio in sulphur dioxide and sulphur trioxide. The results fitted into his original postulate of the Law of Constant Proportions.

With these results, he proceeded to determine the atomic weights of the elements he studied, although at first he had no way of determining whether a given value or some multiple of it represented the true atomic weight. In the Dulong and Petit law, announced in 1819, he saw the tool for his purpose and was able to correct his values. In 1828, he published revised tables of no fewer than 45 elements. Thirty-nine values were determined by him and six by his students.

Chemical symbols

While engaged in this work, Berzelius perceived that the system of full names of the elements was not convenient and ***suggested a substitute, namely the use of the initial of the Latin name or the initial plus the next letter,*** where necessary: O could be written for oxygen, H for hydrogen and Pb for lead (Plumbum). Thus, ***a new international language for chemistry*** was born. Though the use of letters to represent elements was not new, Berzelius introduced a new quantitative concept with his symbols. The chemical formulae of the compounds of these elements represented the chemical proportions of the elements, ***the proportions being indicated by writing 'small numbers as subscripts'*** (for example, SO_2). There was opposition at first to their use, but this system became the international code.

New elements

Much of Berzelius' work involved minerals, which were classified at that time in terms of crystal forms, appearance and physical properties. Finding it unreliable, he devised his own system based on chemical composition. Like many of Berzelius' innovations, his system of mineral classification met with hostility at first. The Royal Society, London recognised this work and awarded him the Copley Medal. His interest in the composition of inorganic substances and of industrial wastes led to ***the discovery of some new elements. He himself discovered selenium and thorium,*** while his students isolated lithium and vanadium.

Isomerism and catalysis

Berzelius is remembered for two important concepts, namely isomerism and catalysis. The former is concerned with compounds of the same chemical composition but having different chemical properties. Isomerism shows that there is more involved in chemical structure than the ratios of the elements and atomic weight. That is, arrangement of atoms determines the chemical properties of a compound.

Catalysis, which he discovered in 1835, is the phenomenon where a reaction occurs only when some third substance was present, although this substance remained unchanged throughout the reaction.

Berzelius never felt at home with organic chemistry, as it involved new principles. "In living nature, the elements seem to obey entirely different laws than they do in the dead", he observed. So his attitude towards this subject became more and more antagonistic, as he grew older.

Imprint on chemistry

As secretary of the Swedish Academy of Science, Berzelius began in 1821 to publish the *Annual Survey of Progress in the Sciences* which he continued till his death (August 7, 1848), by which time 27 volumes had been issued.

Berzellus suffered from poor health throughout his life, the periodic headaches, which occurred regularly each month at the time of the new and full moon. He married at the age of fifty-six (1835). On this occasion, the King of Sweden bestowed on him the title of Baron.

By the skill of his laboratory techniques, by his power to collect, synthesise and publicise the chemistry of his day, and by the group of his Swedish and foreign students who spread his thoughts abroad, ***Berzelius exerted an influence which is reflected in chemistry more than a century after his death.***

Pierre Dulong (1785-1838)

Who founded thermochemistry

Dulong's spectacular discovery was nitrogen chloride. He lost a finger and the sight in one eye, when he tried to prepare it.

Difficult childhood

Pierre Dulong was born on February 12, 1785. His father died shortly after his birth and his mother, when he was four years old. The young boy grew up in his aunt's home at Auxerre where he attended a local college. His mathematical ability was already evident to his teachers. He entered the Ecole Polytechnique, Paris, as a boy of 16.

Excessive study ruined his health and Dulong had to discontinue in the second year of the course. He next turned to medicine, as formal qualifications were not required for admission in the post-revolutionary period. After obtaining the medical degree, he practised in one of the poorer districts and gained popularity. He was so tender-hearted that he not only treated his patients without fee but bore the costs of medicines! So he had to close down the practice.

From medicine to chemistry

After leaving medicine, he turned to chemistry, as this emerging branch of science gave him hope for making a name for himself. The sound mathematical training at the Ecole Polytechnique stood him in good stead in the work he started in physical chemistry. Thus, Dulong literally followed the footsteps of his patron Berthollet who offered him a position in his private laboratory at Arcueil; this enabled him to pursue work in chemical reactions. Besides, he continued with his teaching in local institutions from 1811 to 1813, when he joined the faculty of Ecole Polytechnique.

On the death of his research collaborator Petit (1791-1820), Dulong was offered in 1820 the post of professor of physics, which he accepted, in his own words "through a weakness of character for which I reproach myself incessantly. Even with good health, the duties of this post would have left me with little free time". In 1831, he got relieved from his teaching duties by moving to the post of director of studies, but here he again regretted that he no longer had the laboratory facilities of his earlier post. In his mature years, ***this conflict between his desire to do research and to teach developed into a crisis.*** Dulong was elected to the French Academy of Sciences (1823); in 1828 he became the president for one year. He died in Paris on July 19, 1838.

Study of oxalates

In his first paper from the Society of Aruceil, he provided a striking confirmation of Berthollet's thesis that chemical affinities were not fixed and that ***chemical reactions could often be reversed.*** He showed that when barium sulphate, a notoriously 'insoluble' salt, was boiled with a solution containing an equivalent quantity of potassium carbonate, it was partly decomposed. It was the work on 'insoluble' salts that led him to make a special study of the oxalates of barium, strontium and calcium.

To Dulong, ***a metal oxalate was a simple compound of the metal and carbonic acid,*** though this is contrary to the chemical theory then prevailing. His work contributed to the post-Lavoisier concept of an acid as a compound containing hydrogen that is replaceable by metal; it was thus the forerunner of the radical theory in organic chemistry.

Physical injuries

Dulong's more spectacular discovery (1811) was nitrogen chloride, the spontaneously explosive oil. He lost a finger and the sight in one eye when he tried to prepare it. He suspended the work till October 1812, as he realised that he could avoid an explosion only if he kept the temperature low. After receiving further injuries, Dulong abandoned the investigation.

But the remarks in his memoir are significant: "It seems to me this compound contains a certain amount of combined heat which, when its elements separate from each other, raises their temperature and imparts to them a great elastic force". ***This marked the beginning of thermochemistry.***

Dulong–Petit's law

The interest of his colleagues in the Society at Arcueil prompted him to undertake ***two important lines of research in combining proportions;*** the analyses of phosphoric acid about which there was great divergence between chemists, and the various oxides of nitrogen which had created some confusion in the early 19th century. He collaborated with Berzelius (1779-1848) determining the gravimetric composition of water (1819).

Dulong began in 1815 ***his collaboration with Alexis Petit, a mathematical physicist:*** it resulted in three important memoirs on heat. They were concerned with the specific heats of elements; if these elements existed as atoms, he visualised that there might be a connection between the weight of the atom and the amount of heat required to raise the temperature of a given weight of that element by a certain amount.

Dulong and Petit first had to develop a reliable method of determining specific heats, for the published data were found unreliable. They recorded the specific heats of a dozen metals and sulphur and then multiplied each by the element's atomic weight. This was enunciated as ***a fundamental law that bears their names: Specific heat of an element × its atomic weight = 6.4 (approx).*** Their law of atomic heats was applied to settle disputed values of atomic weights: for example, they halved values given by Berzelius for silver and sulphur. Following his success in relating atomic weights to specific heats, Dulong tried to establish a similar relation between atomic weights of gases and their refractive indexes, but with little success.

Other contributions

Dulong and Petit undertook a complete re-examination of Newton's law of cooling. They broke down a complex phenomenon into its constituent parts, clearly distinguishing between losses due to radiation and losses due to contact with particles of a gas. Thus ***they arrived at a series of laws, which was the starting point of the work of Joseph Stefan (1835-1893).*** It was not until 1879 that Stefan was able to reduce the phenomenon of radiation to a simple law and he acknowledged that his law agreed with the experiments of Dulong and Petit.

Dulong made intuitively generalisations such as: the quantities of heat evolved are approximately the same for the same substances combining at different temperatures, and equal volumes of all gases give out the same quantity of heat.

Applications in engineering

Dulong showed that ***his work on heat had utilitarian applications in engineering.*** He collaborated with Arago (1786-1853), another physicist, on a long and perilous study of the pressure of steam at high temperatures. This work was on account of the French government's concern about the safety of boilers.

(*Dictionary of Scientific Biography,* Vol. IV, Ch. Scribner's Sons, New York.)

JUSTUS VON LIEBIG (1803-1873)

WHO TOOK CHEMISTRY TO STUDENTS

He set the pattern for chemistry education that came to prevail in Germany and steered German chemistry to great heights in the mid-nineteenth century. His analytical methods for research found extensive applications, particularly in the birth of the synthetic dye industry and agricultural chemistry. He turned the results of his research into useful products such as meat extracts and artificial milk.

Passion for chemistry

Justus Liebig was born on May 12, 1803 in Darmstadt, Germany. As a child, he showed a passion for chemistry and was much less interested in the other common subjects in school. His father made family drugs in a small laboratory where young Liebig learned to perform by himself, chemical experiments described in books, which he borrowed from the town library.

Liebig studied under Wilhelm Kastner, a renowned chemist at the University of Erlangen. He received his doctorate degree in 1822. From 1822 to 1824, he studied in Paris. He attended the lectures of Gay-Lussac (1778-1850) and Dulong (1785-1838), which opened up to him a rigorous and quantitative method of learning chemistry, unlike anything he had seen in his country.

He imbibed the traditions of analysts such as Lavoisier (1743-1794) and these ideas were planted in the system of laboratory training that he instituted when he returned to Germany.

Liebig taught chemistry at the University of Giessen from 1825 to 1851. In 1852, he moved to the University of Munich, where he served as professor of chemistry for the rest of his life. Here, he had an institute built to his specifications and with better research facilities than in Geissen and also a large lecture hall.

Birth of organic chemistry

Liebig's scientific reputation grew during 1829-39 when he made many original contributions to ***the systematic formulation of organic chemistry as a discipline and the application of chemistry to biology*** (now known as biochemistry). He set up his laboratory in an unventilated room, with benches along the walls. He had to equip the laboratory and pay the assistant from his salary! ***It was the first laboratory in the world designed to teach young students systematic methods in chemical research.***

Students began by learning ***a system of qualitative analysis***, namely an orderly way to separate and identify the acids and bases present in a solid or solution of unknown composition. Liebig's own practical skill and his magnetic qualities as a teacher attracted to this laboratory students from all over Europe and the U.S. Many famous chemists emerged from this laboratory – August Hoffmann, F. Kekule, Edward Frankland and Wurtz, to mention a few.

After 1838, ***his interest shifted from organic chemistry to the chemistry of plants and animals***. He showed that plants took carbon dioxide, water and ammonia from air and soil; and through the use of mineral fertilisers, the depletion of elements in the soil can be made up.

Formulation of radical theory

His most celebrated investigation was the study of bitter almond oil, in which the same chemical group was found to pass unchanged through a variety of reactions. Such a group was named a radical. The 'radical theory' that he formulated was the first major attempt at systematisation of organic chemistry.

Through his work in organic chemistry, he came to know Woehler (1800-1882). This association turned out to be a lifelong friendship. The book

Organic Chemistry in its Application to Agriculture and Physiology that Liebig published in 1840 was well-received and ran into several editions. He was made a Baron in 1845.

Public lectures

At Munich, Liebig grew weary of the unremitting burden that creative teaching work always carried. So he restrained his teaching activity and took only a few students to work in his laboratory. He ***instituted a series of evening talks to arouse public interest*** on scientific topics. He seemed to have enjoyed the quiet time at the University of Munich more than his earlier 26 hectic years at Giessen, when he was one of the most formidable figures in science. He remained quite active till his death on April 18, 1873.

August von Hofmann (1818-1892)

Who heralded German chemical industry

Despite his pre-eminence as a practical teacher, Hofmann himself was not a good experimentalist. But he had ***an extraordinary acumen to spot unusual skills in his assistants*** who helped perform much of his experimental work. Hofmann led a full and rich life devoted to the pursuit of science. By percept and practice, he showed that chemistry was the most readily applicable science to the creation of wealth.

The 19th century saw Germany as the unrivalled leader in the field of organic chemistry. Justus von Liebig (1803-1873) was responsible for the meteoric rise of German chemistry, for he organised a teaching laboratory where students worked on research projects during their academics and received intensive training in industrial applications of chemistry. One of his students was August von Hofmann (1818-92) who continued this method of ***teaching chemistry through laboratory instruction*** and introduced it in England later.

Influence of Liebig's lectures

Hofmann was born on April 8, 1818 in the small university town of Geissen in Germany. He joined the local university in 1836, electing to study law and languages. But ***attracted by Liebig's chemistry classes,*** he turned to the study of chemistry.

Under Liebig, Hofmann had to carry out ***research on coal tar, the waste product of coal gas industry.*** From such raw material often dumped in rivers, he extracted basic oils by titration with hydrochloric acid. For his investigations into these organic bases, he received in 1841 his doctorate. His thesis *'Chemical investigation of the organic bases in coal tar'* published in 1843, laid the foundation for his life's work.

Career in England

On an invitation from Queen Victoria, he accepted the appointment as Director of the Royal College of Chemistry and moved to London in 1845. Despite the financial insecurity of the college, which was started as a private school, Hofmann continued to work in England for 20 years. He trained a generation of brilliant chemists, notably W.H. Perkin, C.B. Mansfield, William Crookes, Peter Griess and Alexander Martius. The course of lectures that he had given during this period was published as a book titled *An Introduction to Modern Chemistry.* A special edition of his book was brought out on the occasion of his bidding farewell to his many British friends.

Professor of chemistry at Berlin (1865-1892)

Hofmann returned to Germany in 1864. He was elected in 1865 to the Chair of Chemistry at the University of Berlin, a post he held with distinction until his death on May 2, 1892.

Hofmann played a leading role in the growth of nitrogen containing compounds then known by the name aniline in the chemical world, He used aniline from indigo in substitution reactions that introduced chlorine and tested the properties of the products. Since the study of synthetic methods was becoming fashionable, he tried laboratory synthesis of aniline from coal tar. This he could achieve in two steps — nitration and reduction.

A result of great simplicity

On May 18, 1863, he established that aniline blue was rosaniline into which three phenyls had been introduced. It was acclaimed as a result of great simplicity ***which contributed to deriving useful colours from other substituted products of aniline red.*** Alkylation reaction was immediately undertaken with ethyl iodide and a blue colour was obtained. The colour changed to violet with more iodide. The new colourants were an immediate success, by which Hofmann made a considerable fortune.

In collaboration with his teacher Liebig, Hofmann attempted to produce a substitute for quinine, a chemical base like aniline. But he drew a blank. But ***this failure inspired W.H. Perkin (1838-1907) to pursue with the experiments that eventually led to aniline purple or 'mauve',*** which was obtained by oxidising aniline with potassium dichromate. This heralded ***the beginning of the synthetic dyestuffs industry in 1857.*** Hofmann had laid the foundation of this industry.

Hoffmann's reaction

Hofmann's other significant accomplishments were:

- production of formaldehyde by passing vapours of methyl alcohol over hot platinum,
- the reaction, named after him, for converting an amide into amine,
- solution to the problem of substituting atoms of chlorine for the hydrogen atoms in the aromatic compounds,
- method of determining the molecular weights of liquids from vapour density and
- ***modern systematic names to the aliphatic hydrocarbons.***

Hofmann published 360 major papers, which grew out of his work with the development of coal tar, and the many difficult synthesis of organic compounds.

Honours

He received all the honours in his country: he was elevated to peerage in 1888. He served for fourteen years as President of the German Chemical Society.

FRIEDERICH KEKULE (1829-1896)

WHO CLASSIFIED ORGANIC COMPOUNDS

How the now familiar ring structure emerged is better told in Kekule's own words: "There I sat and wrote my 'Lehrbuch', but it did not proceed well. I turned the chair to the fair place and fell half asleep. My mind's eyes, trained by visions of a similar kind, now distinguished larger formations of various shapes. Long rows, in many ways more densely joined, everything in movement, winding and turning like snakes. And look, what was that? ***One snake grabbed its own tail and mockingly, the shape whirled before my eyes".***

Influence of Liebig's course

Friederich Kekule was born on September 7, 1829 in Darmstadt, Germany. He attended the local gymnasium where he distinguished himself by his studiousness and talent for drawing. Obedient to his father's wishes, he enrolled in 1847 in the school of architecture at the University of Geissen. During the second semester, ***an enjoyable course in chemistry under Liebig made him switch from architecture to chemistry.***

In the winter of 1850, he began to work in the laboratory of Liebig. On the latter's advice, he proceeded in 1851 to Paris and took courses in physics and chemistry. This stay proved to be rewarding, as he became the student (and later his friend) of the eminent chemist Charles Gerhardt (1816-1856).

He was one of the distinguished students of Justus von Liebig (1803-1873) who propounded the theory of the ability of carbon to form a vast array of compounds and its unique self-linking capacity.

On returning to his country, he was awarded in June 1852 the doctorate for his thesis on *The ester of anyl-sulftiric acid.* From 1853 to 1855, he worked in London with William Williamson whose work influenced the development of Kekule's theoretical ideas.

Kekule was resolute on pursuing a university career; in 1856, he began teaching organic chemistry at the University of Heidelberg. Like his mentor Liebig, ***he installed at his expense a lecture room and private laboratory*** where he carried out his experiments on the chemical constitution of fulminate of mercury.

Rise of organic chemistry studies

In 1858, he was invited by the University of Ghent to revivify the teaching of chemistry in Belgium and was Professor till 1865, when he moved to Bonn. He continued his scientific career at the University of Bonn — rejecting an invitation from the University of Munich to be the successor to Liebig — until his death on July 13, 1896. During his long tenure, he contributed to the spectacular rise of organic chemistry in Germany. His students came from all over Europe.

Work on theoretical chemistry

Kekule was not a skilled experimentalist but ***an inspired pedagogue. His mind was keyed to theoretical problems.*** He drew structural formulae for organic molecules and showed that the atoms of carbon can link together in long chains, which then can form various polyvalent radicals. An ordered classification of organic compounds thus became possible.

Honours

In 1868, Kekule was awarded the Ph.D. from the University of Bonn, in recognition of his contributions to theoretical chemistry. He wrote a four-volume textbook of organic chemistry. When he was ennobled for his scientific achievements, he added 'von Stradoniz' to his name.

Obscurity of Couper's work

This idea was announced almost simultaneously (in June 1858), but independently by the chemist A.S. Couper (1831-1892) from Edinburgh. On the basis of his publication in May 1858 in *Liebig's Annalen*, Kekule claimed that priority must go to the person who first enunciated his ideas. Sadly, Couper was denied priority by so small a margin: while Kekule was destined to achieve world fame, Couper sank into total obscurity and lived the last thirty years of his life mentally disturbed, cared for only by his mother.

Robert Bunsen (1811-1889)

Of burner fame

Bunsen was the first to produce magnesium in the metallic state. He demonstrated the brilliance and reaction-producing qualities of the flame, when magnesium is burnt in air.

Early education and career

Bunsen was born on March 31, 1811, at Gottingen (Germany), where his father was the chief librarian and professor of modern languages at the University. He attended the local school and continued his studies further at the gymnasium at Holzmuiden. He entered the University of Gottingen in 1832 and studied chemistry, physics, mathematics and mineralogy. Bunsen received his Ph.D in chemistry in 1833.

Bunsen toured Europe during 1830-1833 on a Government grant, visiting laboratories, factories and places of ecological interest. In particular, he visited Liebig (1803-1873) in Geissen, Gay-Lussac (1778-1850) in Paris and Mitscherlich (1794-1863) in Bonn. In the fall of 1833, Bunsen became an honorary teaching assistant at Gottingen. From 1838, he taught at the Universities of Marburg, Breslau and elsewhere. At Breslau, he came to know Kirchhoff (1824-1887) with whom he later did important research in spectroscopy.

A renowned school of chemistry at Heidelberg

In 1842 Bunsen was appointed professor of chemistry at the University of Heidelberg where he remained till his retirement in 1889 at the age of 78. Bunsen did his research in a laboratory specially constructed by the Government of Baden. Here he guided the work of numerous young men and built up a famous school of chemistry. Bunsen never married and lived for his students: teaching and research occupied most of his time.

In 1837 he began his only venture into organic chemistry with an investigation of the highly toxic arsenic-containing compound cacodyl. His six-years' work resulted in five papers published between 1837 and 1842. He found an antidote to arsenic poisoning in freshly precipitated, hydrated ferric oxide. He lost sight in one eye, from an explosion of cacodyl. Eventually, ***he barred research work in organic chemistry*** in his laboratory.

Between 1838 and 1846, Bunsen investigated the composition of gases from blast furnaces in Germany and England and showed that 50–80 per cent of the heat was lost. ***He formulated recycling techniques for efficient utilisation of the heat generated by the fuel*** and further established that valuable by-products such as ammonia could be retrieved.

Bunsen compiled all these results of his research in his only book *Gasometric methods* (1857), which dealt with techniques of audiometric analysis. This work brought analysis to a level of accuracy attained earlier by gravimetric and titrimetric techniques.

Between 1840 and 1850, Bunsen made a number of improvements in the galvanic battery. In 1841, he invented the carbon–zinc electric cell now known as the Bunsen battery. Bunsen was the first to obtain magnesium in the metallic state from the fused chloride. He demonstrated the brilliance and reaction-producing qualities of the flame when magnesium is burned in air.

Bunsen burner

Bunsen improved the burner invented by Aime Argand and Michael Faraday to produce a hotter flame than that was obtained previously, using ambient air and gas alone. The Bunsen burner, with its non-luminous flame, quickly attained a unique position in ***the dry tests of analytical chemistry.*** It made it easy ***to identify metals and their salts by their characteristic coloured flames.***

Other contributions

In the 1860s, Bunsen and Kirchhoff (1824-1887) collaborated to develop the field of spectroscopy. This led to ***their announcement (1860) of a new alkali metal, cesium,*** detected in a few drops of alkaline residue. Cesium salts had earlier been mistaken for compounds of potassium. Bunsen succeeded (1862) in isolating metallic rubidium. Several other elements identified by other eminent workers, using spectroscopic methods were thallium (1861), indium (1863), gallium (1875), and germanium (1879).

Between 1852 and 1862, Bunsen worked with Roscoe (1833-1915) on petrochemical research involving the chemical combination of equal volumes of hydrogen and chlorine under incident light. The effect was related to the wavelength and followed a law of inverse squares. ***Bunsen and Roscoe were founders of a whole new science, photochemistry.***

Work in geology

A man of wide interests, Bunsen worked extensively in geology and mineralogy. He performed chemical analysis of eruptive rocks, insisting that instead of determining the mineral composition, the chemical composition of the rock in its entirety should be ascertained. He concluded that volcanic rocks are mixtures of two extreme kinds of rock, acidic and basic. Although this explanation is no longer accepted, his observations contributed to the development of modern petrology.

In the mid 1850s, Bunsen with his co-workers, ***isolated lithium and several alkaline earth metals, barium and calcium;*** he also prepared the rare earth metals of the cesium group. To obtain the specific heats of these rare elements, Bunsen devised a sensitive ice calorimeter that measured the volume rather than the mass of ice melted, thus requiring only a small sample of the metal. From the specific heats, the atomic weights of these elements were readily obtained.

A devoted teacher

As professor at University of Heidelberg, he built a school of chemistry. He was a most devoted teacher. He delivered every year 100 lectures covering a course 'General experimental chemistry'; the contents changed little through the years. Organic chemistry was excluded; neither the Mendeleev's periodic table nor Avogadro's hypothesis found a place, because he emphasised the experimental side of science. He was a skilled glassblower and crystallographer.

Bunsen developed and improved several pieces of important laboratory equipment, which included the ice calorimeter, steam calorimeter, filter pump, thermopile, galvanic battery and the Bunsen burner. The latter was widely used to identify metals and their salts by the characteristic colour of the flame. ***Spectrum analysis thus gained great importance in the study of the sun and the stars.***

Honours

Bunsen received several honours — FRS (1858), member of the Academy of Sciences, Paris (1882), Copley Medal (1860), Davy Medal (1877). He died on August 16, 1899 at Heidelberg.

(*Dictionary of Scientific Biography,* Vol. II, Ch. Scribners Sons, New York)

DMITRI MENDELEEV (1834-1907)

WHO GROUPED THE ELEMENTS

Mendeleev's insight into the similarities of different elements led him to develop ***the Periodic Table.*** This provided a vital link between chemistry and all other branches of fundamental sciences.

Traumatic boyhood

Dmitri Mendeleev was born on February 8, 1834 in the Siberian town of Tobolsk, as the seventeenth and last child in the family. He passed through a traumatic period in his boyhood. His father, the principal of a local high school, went blind suddenly. To support the family and provide Mendeleev higher education, his mother had to fight against heavy odds. She went to St. Petersburg and got young Mendeleev admitted into a college. Her mission accomplished, she died soon after.

Dmitri Mendeleev ***displayed brilliant intellectual ability, sharp memory and fascination for mathematics, physics and history.*** He graduated from the University of St. Petersburg in 1855, and started his research under A. A Voskresenkil, the grandfather of Russian chemistry.

His career

During 1859-60, Mendeleev was with the University of Heidelberg where he first collaborated with Bunsen (1811-1899) and then established his own laboratory. In February 1861, he returned to St. Petersburg and resumed teaching chemistry. In 1865, Mendeleev defended his thesis for the doctorate

'On the compound of alcohol with water'. He first developed the view that solutions are chemical compounds and that dissolving one substance in another is not to be distinguished from other forms of chemical combination.

A turning point in his career occurred in 1867 when he was appointed to the chair of inorganic chemistry at the University of St. Petersburg, which he retained for the 23 years that followed. In preparing for his lectures, he found nothing which he could recommend to his students as a textbook. He therefore set out to write his own book *Principles of Chemistry,* which was highly successful.

Due to some student disturbance, his teaching career came to an abrupt end in April 1890. From 1892 he worked on the regulation of the system of weights and measures and introduced in 1899 the metric system in Russia. He held the post of Director, Central Board of Weights and Measures, until his death on February 2, 1907.

The Periodic Table

It is difficult to determine when Mendeleev first hit upon the periodic table; ***while working on his book, he was impressed with certain irregularities of the chemical properties of elements.*** So on March 18, 1869, he presented at the meeting of the Russian Chemical Society his paper 'An outline of the system of elements based on their atomic weights and chemical similarities'. ***His table hardly attracted any attention, until his predictions were fulfilled by the discoveries of gallium (1874) scandium and germanium (1885).*** The major drawbacks pointed out at that time were the difficulty in accommodating the rare earth group and the chemically inert elements.

In recognition of this contribution, which had won for him ***an immortal place in science,*** his name was proposed in 1880 to the Russian Academy of Sciences but was defeated. Leo Tolstoy contributed to his defeat and found in Mendeleev a formidable opponent, when the former sought to limit the teaching of science in schools. For, Mendeleev saw in science the essential ingredients of the educated mind.

Students followed his funeral to the cemetery in St Petersburg and carried ***the Periodic Table*** of the elements high above the procession as the ***fitting emblem of Mendeleev's scientific career.***

William Perkin (1838-1907)

Who discovered synthetic dyes

Perkin's bichromate method gave impetus to the birth of a new waltar dye stuff industry. He retired from industrial career at the age of 36 and switched to a career of private research in pure chemistry.

Influence of Hofmann's lectures

William Perkin was born in London on March 12, 1838 as the son of a building contractor. Perkin had his early education in the city of London. As science was not a part of the school curriculum, ***he attended weekly classes in chemistry held by one of the masters during dinner,*** which aroused his interest in chemistry.

Perkin's father was opposed to his making a career as a chemist; but his master, by name Thomas Hall, fortunately intervened. He got the young boy enrolled in 1853 in the Royal College of Science. Here he attended the lectures of the famous German chemist August von Hoffmann (1818-92), who had established his original method of teaching chemistry through laboratory instruction. At the end of his second year in the college, Perkin was appointed Hofmann's assistant. Simultaneously, he set up his own laboratory to do private research.

The first dye stuff from coal tar

In collaboration with his colleague A.H. Church, he obtained one of the first azo dyes derived from naphthalene and patented the process. ***Perkin's major discovery, that of mauve, followed thereafter*** (1856). While Perkin was working in Hofmann's laboratory, he undertook the synthesis of quinine. Pursuing his own line of research, Perkin obtained not quinine but a dirty, dark-coloured precipitate. He followed his intuition and proceeded to examine this precipitate further, when he discovered its excellent colouring property. Thus, he isolated aniline purple that later came to be known as mauve, which is ***the first dye to be produced commercially from coal tar,*** the waste product of coal gas in industry. He patented his technique.

Birth of a new industry

With the financial support given by his father, Perkin began constructing a factory at Greenford Green in June 1857. He had not only to work out a new technique for converting nitrobenzene to aniline but also to devise a new apparatus. Perkin's 'bichromate method' gave impetus to ***the birth of a new coal tar dye stuff industry.*** Also he discovered a chemical process for preparing unsaturated acids and he used this process to produce coumarin, the first artificial perfume.

In 1868, Perkin worked out two new methods to manufacture alizarin more economically, both using coal tar products. Synthetic alizarin soon replaced 'rose madder' as the prime red dye. By 1971, Perkin's factory was manufacturing 220 tons of alizarin a year.

Pursuit of science

Perkin had always cherished the ambition to devote himself to pure science. He took the first step in 1873; he sold his factory and patents, and ***retired from an industrial career, when he was 36.*** Then he ***switched to a full-time career of private research*** in pure chemistry. This demanded great courage and determination in the pursuit of science.

A new branch of science

Even while burdened with commercial responsibilities in the dye stuffs industry, he made significant contributions to organic chemistry. In particular, his discovery that cinnamic acid could be synthesised from benzaldehyde led to the first synthesis of indigo by Baeyer and Cavo. Perkin was accorded ***more recognition for his work by scientific workers in pure chemistry*** than by his co-workers in the field of dye manufacture. Jubilee celebrations were held in 1906 in England and the U.S. Perkin was knighted on this occasion.

Perkin's devotion to science was that he continued to work until the time of his death (July 14, 1907). By his investigations, the property of the magnetic rotary polarisation of organic compounds became important in the study of molecular structure.

Mr. Bruehl, one of the pioneers in the field, wrote in 1906 to Perkin commending his work: 'Before you began work there was little, almost nothing, known of this subject, certainly nothing of practical use to the chemist. ***You created a new branch of science'.***

(*Dictionary of scientific biography,* Charles Scribner & Sons, New York.)

William Ramsay (1852-1916)

Who discovered inert gases

After a period of combined research with Lord Rayleigh, he was convinced that there was an entire group of elements missing from the Periodic Table. He embarked on a diligent search with his student Morris Travers.

Early life

William Ramsay was born on October 2, 1852 at Glasgow, Scotland, in a family noted for scientific attainment. After completing his secondary education at the Glasgow academy, he studied classics, mathematics and literature in the university. He attended in 1870 the physics lectures of William Thomson (1824-1908), known as Lord Kelvin; simultaneously ***he worked for 18 months as an apprentice chemist.*** Though originally intended for the clergy, his latent interests in science gradually blossomed.

In 1870 Ramsay became a student of the famous German chemist Robert Bunsen (1811-1899) at the University of Heidelberg; but after a year, he went to the University of Tubingen to work under Professor Fittig. Ramsay did research on nitrotoluic acids and received his doctorate in August 1872 at the age of 19.

Professor at Bristol and London

On his return from Germany, Ramsay joined in 1874 as a tutorial assistant at the University of Glasgow and in 1880 became professor of chemistry at the University College, Bristol. He shifted in 1887 to the University of London, where he was appointed to the Chair of Chemistry. He served for 25 years till 1912, when he retired to Hazelwood. He built a private laboratory and was active in scientific pursuits till his death on July 23, 1916.

At Glasgow, Ramsay worked in the area of organic chemistry. In Bristol, he changed to physical chemistry and demonstrated the complexity of the molecular structure of pure liquids. With his assistant Sydney Young, it became his main line of research from 1892. ***They published more than 30 papers on vapour pressure and critical states of liquids.*** This collaboration continued even after Ramsay shifted to London University. Ramsay improved his laboratory techniques, learned the art of glass blowing and laid ***the foundation for his later research on gases.***

Combined research with Rayleigh

In London, Ramsay gradually shifted his investigations to the various aspects of gas analysis such as the accurate determination of the density of gases. After the announcement of Rayleigh in September 1892 on the small difference between the density of atmospheric nitrogen and that of chemically pure nitrogen (first noted in 1785 by Cavendish), Ramsay speculated upon its cause.

Concerning this confirmation of the discrepancy, Ramsay recalled in a letter (August 1894) that "Rayleigh gave numbers about which there could be no reasonable doubt. I asked him then if he minded my trying to solve the mystery. He thought that the cause of the discrepancy was a light gas in non-atmospheric nitrogen; I thought that the cause was a heavy gas in atmospheric nitrogen. Rayleigh spent the summer looking for the light gas; I spent it hunting for the heavy one. And I have succeeded in isolating it" (*Life of Ramsay*, Morris Travers, London University Press, 1956).

Isolating argon

During a further period of combined research, Ramsay and Rayleigh named the new element argon, comprising nearly one per cent of the atmosphere; determined that it is monatomic and found that it has an atomic weight of 40.

Ramsay, now convinced that there was an entire group of elements missing from the periodic table, embarked on a diligent search for them.

In 1898, with the collaboration of Morris Travers (1872-1961), Ramsay confirmed the presence of ***krypton, neon and xenon in the atmosphere.*** Neon, being more volatile, resisted liquefaction by means of liquid-air techniques. Therefore Travers rigged up a liquid hydrogen apparatus based on Dewar's recent demonstration of the feasibility of obtaining adequate quantities of liquid hydrogen. They were thus able to solidify neon and by July 1900 they had determined its properties.

Promoting educational excellence

The series of books Ramsay published was the culmination of a lifetime of work and devotion to the cause of science and education: A *system of inorganic chemistry (1891), The gases of the atoms* (1896), *Modern chemistry (1900), Introduction to the study of physical chemistry* (1904), *Elements and electrons* (1913). He was the author of nearly 300 papers.

Ramsay was ***an outstanding and patient teacher.*** He was active in the pursuit of educational ideologies and tirelessly worked for granting independent status to provincial colleges. His efforts were rewarded in 1809, with both Bristol and Owen Colleges being raised to University status. He promoted educational excellence by ***making research an integral component of some select science courses at the University of London.***

Honours

Ramsay was made a FRS in 1888; in 1902 he was knighted; and in 1904 he received ***the Nobel Prize for Chemistry.*** He earned unique fame as the discover of an entire family (5) of inert gases. Colleagues, students and acquaintances always spoke of ***his boyish vigour, total involvement in scientific work*** and personal charm which remained throughout his life. Ramsay was a great general who wanted an able chief-of-staff. "All his best work was done with a colleague". *(Letter from Travers to Soddy,* October 10, 1949)

Svante Arrhenius (1859-1927)

Pioneer of modern physical chemistry

Arrhenius' idea that the molecules of a solute were dissociated even in the absence of a potential difference seemed to many chemists wild.

Doctoral thesis

Svante Arrhenius was born on February 19, 1859 at Uppsala, Sweden, in a family of farmers. He joined Uppsala University, intending to study chemistry.

As the subject was badly taught, he changed to physics. He wanted to pursue research, but it was then possible at Uppsala to do research only on spectral analysis. So Arrhenius shifted in 1881 to Stockholm to do research on solutions and electrolytes.

In June 1883, he presented the results in ***his thesis, which fetched only a fourth class, and was not sufficient to qualify him for an academic position.***

Theory of electrolytic dissociation

Arrhenius was bitterly disappointed. He sent copies of his thesis to eminent chemists such as Clausius, Ostwald and van't Hoff who were impressed with his work.

His thesis contained the germ of the theory of 'dissociation', which term he did not use; but he applied the concept to explain double decomposition.

When an electric current passed through a solution of an electrolyte, the molecules were decomposed. This had been long recognised, but the idea that the molecules of the solute were dissociated even in the absence of a potential difference, seemed to many chemists wild.

From 1883 to 1888, he went around Europe to enlist support for his proposed theory.

The crucial paper, which confirmed his thought, was that of van't Hoff (1886): the latter showed that electrolytes behaved in osmotic pressure experiments, as though there were more molecules in solution than what was the case.

Icy reception

In 1887, the young Savante Arrhenius got an icy reception when he proposed that neutral compounds such as hydrogen bromide (HBr) could completely dissociate in water, releasing independent, oppositely charged ions. ***Three decades later, Max Born showed the ionic dissociation was accompanied by release of substantial energy,*** when an ion is immersed in a dielectric (*Science,* 4 October 2002).

In 1891, Arrhenius was offered the post of professor at Geisen, but he declined on patriotic grounds. He joined the 'Hoegskola' at Stockholm (Royal Institute of Technology), which reluctantly made him a professor in 1895.

Arrhenius was appointed in 1905 a director of the newly founded Nobel Institute at Stockholm, a post that he held to the end of his life. Here he enjoyed ideal conditions for research and writing. He died on October 2, 1927.

Honours

He was awarded the Davy Medal of the Royal Society, London in the year 1902, Gibbs Medal in the year 1911, Faraday Medal in the year 1914. Arrhenius received ***the 1903 Nobel Prize for Chemistry.***

Popularising science

He devoted his later years to the popularisation of science. His books and articles on the treatment of the problems of technical chemistry met with worldwide success. They were translated into several languages and went through numerous editions.

Morris Travers (1872-1961)

Who discovered the missing elements

His record as a scientist and an academic administrator was enviable and difficult to match. He wrote a lucid biography of William Ramsay and a beautiful account of the discovery of the rare gases, in his eighties.

Research under William Ramsay

Morris Travers was born in Kensington, London, on January 24, 1872, as the second son of an eminent surgeon at the Charing Cross Hospital. At school, he won many prizes and showed an unusual interest for science at that tender age, which remained till his death in 1961.

At the University College, London, where Sir William Ramsay (1852-1916) was professor, ***Travers attended lectures on a multitude of subjects — physics, chemistry, pure and applied mathematics, zoology, botany and physiology.*** He found that while the laboratory training Ramsay gave was quite unsystematic, it was highly personal. The students could actually interact with the great scientist.

Ramsay noticed Travers' experimental skills and appointed him as one of his three personal demonstrators. The professor, as was the custom then, had to pay their salaries (£50 per annum each). A worse convention was that if, by chance, the demonstrator had any inclination to do original research — as was the case with Travers — he had to meet all his research expenses from his own small stipend!

Discoveries of missing elements

Travers began studying in 1894 the properties of the new element argon, which Ramsay (along with Lord Rayleigh) had just discovered. Theirs was a great and ***ideal collaboration "in the search for missing element or elements which the periodic law indicated should exist".*** Their first paper was entitled 'Helium, a constituent of certain minerals'. Later came the discovery of krypton, xenon and neon (in that order) the arduous work of separating the gases using liquefaction techniques and showing them to be monatomic.

Travers was elected to the Royal Society at the early age of 32. He discovered a method of preparing acetylene from calcium carbide, anticipating Moissan, who is usually given credit for this. Important industries were established in Great Britain on the basis of this work. He did such work for the development and production of scientific glassware. He was involved in the design of many chemical engineering plants and the fabrication of small and large furnaces, especially for the glass and ceramic industry.

As an experimental scientist

His work on the thermal decomposition of organic vapours, production of water gas and coal gassification attracted much attention. He was the recipient of the medal of the Institute of Fuel. For his contributions to the glass industry, he was elected Honorary Fellow of the Society of Glass Technologists, of which he became the president. He was also the president of the Faraday Society (1936-38).

In 1956 Travers wrote a lucid biography of his mentor Sir William Ramsay and a beautiful account of the *Discovery of the Rare Gases.* Christopher Ingold described this book as "enthralling.... vivid, cogently documented, thorough and technically precise". He ended by saying, ***"In this volume, written by Professor Travers between his eighty-first and eight-third birthdays,*** with all the strength of a man in his prime, ***we find a remarkable combination of high literary merit, good history and excellent science".***

Travers excelled as an experimental scientist; he had an aptitude for applied science, a profound knowledge of university administration, and had done yeoman service to the Bristol University. ***Ramsay suggested that Travers take up the directorship for the newly founded institution in India*** (Bangalore), which had not yet been named, "consequent on the bequest of the late J.N. Tata of Bombay". He was to start a nucleus of scientific and technical departments, which would develop along the lines of the Imperial College of Science and Technology, London. He was offered a salary of £1800 p.a., a free house and other perquisites. His salary in England was £350 p.a. and "out of this he had to meet considerable expenses in connection with his researches". Even so, he was reluctant to take up the Indian position but was finally persuaded to accept.

First Director of Indian Institute of Science (1906-14)

From 1906 to 1914 Travers was in India. Travers was a very remarkable man, scientist and engineer. ***His record was enviable and difficult to match; the discovery of three new elements,*** applications of many aspects of chemistry to industry, the strengthening of a university in England, and ***laying the ground rules for a research institution in India.*** Travers did an excellent job starting with the layout of the Institute followed by designing and getting constructed many of its major buildings and laboratories. Over the years, the Institute developed along the lines he had laid down. Much credit is due to him for "his wisdom, farsightedness and thoroughness in planning".

In England, he was described as a very capable administrator, who liked things done quickly and was not tolerant of inefficiency. He lived up to this description even in Bangalore. However, in India, he was described as a person wanting in administrative skills and who did not have the tact needed for the direction of an institution, the first of its kind in the country.

Production of liquid hydrogen

James Dewar (1842-1923) claimed that he produced liquid hydrogen in 1895 but no description of the apparatus or the procedure was ever published. Travers set out to liquefy hydrogen in quantities required for real experiments. He modified a small air liquefier and the apparatus operated successfully and produced enough liquid hydrogen to fractionate a helium–argon mixture. Legend has it that the air liquefier, which was in the central workshop of the Institute, was set up by Travers. C.V. Raman was often seen watching the kettle impatiently for the production of liquid air by this machine for his spectroscopic and X-ray studies.

In 1908, Travers took his liquefier to Berlin and produced liquid hydrogen for the first time in Germany. This original and historic hydrogen liquefier is now in the Science Museum, South Kensington. The one which probably produced liquid hydrogen in India for the first time (circa 1910) ended up, perhaps, in the scrap heap.

A memorial lecture

Few in India know that it is at the I.I.Sc (1911-14) that Travers (along with N.M. Gupta and R.C. Ray) did his pioneering work on boron and its compounds. The Council of the Indian Institute of Science, Bangalore, instituted in 1992, the 'Morris Travers Memorial Lecture' in memory of its first director.

Yes, Travers was quite a remarkable man, always excited about science whether it was done by him or by others. To him every new result was a discovery, something exciting, to be talked about and discussed. He was feverishly active all the time and never relaxed except when he was in his home with his family or with his music.

IRENE JOLIOT-CURIE (1897-1956)

DISCOVERER OF ARTIFICIAL RADIOACTIVITY

She had inherited the expertise, single-mindedness and scientific insight of her parents. Her doctoral thesis was a study of the alpha rays from polonium.

How she learned science

Irene Curie was born on September 12, 1897 at Paris, in a family of dedicated research workers. ***She did not attend school until the age of twelve,*** but in 1907 her mother took charge of her scientific education by sending her to the 'teaching cooperative' established by Marie Curie (1867-1934) and her friends for their children. Here Marie Curie taught physics, Paul Langevin mathematics, and Jean Perin, chemistry. This private education lasted till 1909.

Her grandfather, Eugene Curie, who was living with the family, exerted a great influence on Irene's personality. He was the source of attachment to the liberal socialism to which she remained faithful all her life.

She next went to College Sevigne where she received a Bachelor's degree in 1914. She then attended the Sorbonne University (1914-1920) and obtained the 'licence' in physics and mathematics.

Assisting her mother

During World War I, she served for many months as an army nurse and assisted her mother in installing the equipment for the radiography of the wounded.

In 1918, Irene Curie took up the post of an assistant at the Radium Institute headed by her mother. ***She began research in classic radioactivity;*** she conducted a series of studies on the fluctuations in the range of alpha rays. She determined these variations by photographing the tracks that the rays formed in a Wilson cloud chamber. These results were presented in her doctoral thesis in 1925.

Irene married Fredric Joliot in October in the year 1926. From 1931, they worked with alpha particles for two years, and in 1933-34 they observed that aluminium continued to emit positrons even after the bombardment with alpha particles had ceased. They concluded, rightly, that ***a new unstable isotope*** (of phosphorus in this case) ***had been produced in the laboratory*** — the discovery of artificial radioactivity. For this work and its development, they were jointly awarded the ***Nobel Prize in Chemistry in 1935***.

In their Nobel lectures, Fredric the physicist, dealt with chemical identification of the artificially created radioisotopes; Irene the chemist, elaborated on the discovery of a new type of radioactivity, namely the positive beta decay. Marie Curie, who died in July 1934, was not destined to witness the triumph of her daughter and son-in-law.

Irene Joliot-Curie was elected Professor at Sorbonne in 1937. She, however, continued to work at the Radium Institute of which she became the director in 1946.

Discovery of fission

It was during these years that Irene did her most valuable individual work. With her specialised knowledge of radiochemistry, she sought to analyse the complex phenomena that result from bombarding uranium with neutrons.

Otto Hahn and Lise Mietner subsequently repeated these experiments. They proved that ***the neutrons could induce the bipartition of the uranium atom into two atoms of a comparable mass, which was soon afterward termed 'fission'.*** Irene had thus instigated this celebrated discovery.

From 1946 to 1950, she and her husband were directors of the French Atomic Energy Commission. ***Although devoted to science, she loved French***

and English literature. She found great joy in motherhood and devoted much time to rear her children till adolescence.

Victim of radioactivity

She was an inadequately protected nurse during her service earlier in the army and then in the laboratory, when the dangers of radioactivity were not fully realised. She died at the age of 59 (March 17, 1956) a victim, like her mother, of acute leukemia.

(*The Dictionary of Scientific Biography*)

DOROTHY HODGKIN (1910-1994)

CHEMISTRY'S CRYSTAL SEER

Penicillin was discovered in 1928 but its structure, namely the arrangement of atoms in the crystal, though derived earlier by purely chemical methods, could be determined only in 1945. It is Dorothy Hodgkin who ***extended the application of X-ray crystallography,*** which was already used by William and Lawrence Bragg to determine the simple cubic structure of sodium chloride crystals.

The experiment using X-rays to obtain diffraction patterns is straightforward but the analysis of the results was laborious and time consuming. Hodgkin pioneered the use of computers to handle these complex mathematical analyses. In fact, the advances in computer technology made a large contribution to her work 'on the structure of biochemical compounds essential to the understanding and control of pernicious anemia'.

Chemistry kit as a present

Dorothy Hodgkin was born on 12 May 1910 in Cairo, where her father served in the British Education Service. Her interest in chemistry was sparked off by a family friend, Dr. Joseph, who presented her with a chemistry kit. Young Hodgkin then started conducting experiments in her own mini-laboratory. She did exceedingly well in school. Her father lined her up for higher education at Oxford. She recalls "it was then extremely unusual for a woman to read science. We had to be chaperoned... and be back in our rooms by nine o'clock". She qualified at the stiff entrance examination and joined the Somverville College in October 1928 for a degree in chemistry.

Advantage of research in academics

The chemistry course included one year in compulsory research, so the degree took four years instead of three — to her advantage, as she could begin her research career in her undergraduate course. She received intensive training in X-ray crystallography, under H.M. Powell, which involved gathering, and then analysing experimental data. After a brief stint at the Cambridge University, Hodgkin returned to Oxford in 1934.

From then onwards, ***she leapfrogged the normal path — from undergraduate student to the rank of Professor.*** In the process, she completed her doctoral thesis on her own and carried out her Nobel Prize-winning work as an independent scientist. All she needed for her work was a couple of goniometers. She retained her position as a tutor in college and enjoyed teaching.

The technique of X-ray diffraction was developed after World War I. When she adopted this technique to determine the structure of penicillin and finally succeeded in 1945, the process was far more labour intensive.

Structure of insulin

Hodgkin had her first encounter with computers in 1946, when she visited the U.S. Computer analysis made a large contribution to her research work which bore fruit in 1956, after eight years' labour, in the discovery of the structure of vitamin B12 (more than 90 atoms). She had to send her results by surface mail to Cal Tech in the U.S. for computer analysis. Computer technology eventually caught up with Hodgkin's research to produce electron density maps. Thus in 1969, ***she determined the structure of insulin,*** a compound that contains more than 800 atoms, ***which she had started 30 years earlier.***

Honours

Hodgkin received all the honours a scientist could aspire for: FRS (1947), Royal Society research professor (1960), Nobel Prize (1964) and Order of Merit (1965) (the second woman to this distinction since Florence Nightingale).

Her vision of how chemistry and crystallography might combine to reveal the biological functioning of complex molecules eventually unravelled the structures of vitamin B12, insulin and penicillin.

Tribute

One of the colleagues in Hodgkin's laboratory pays a tribute to her superior style of research: ***"She was very intuitive in her science.*** She had clear objectives when she chose to study the crystals; ***she believed that their biological function would be revealed through their crystal structure.*** Each was a challenge crystallographically; she thought it could be done, and she was willing to persevere for years". (*New Scientist,* 23 May 1992.)

Pavlov

SECTION FIVE

Darwin

LIFE SCIENCE

G. N. Ramachandran

WILLIAM HARVEY (1578-1657)

FOUNDER OF MODERN EXPERIMENTAL PHYSIOLOGY

Living matter became an object of scientific enquiry only in the seventeenth century. It may be that the art of healing which led to the sciences of anatomy, physiology and biology assumed importance when it was felt life could not be effectively aided without such knowledge.

Obviously, the complexity of all processes taking place in living matter was sufficiently alarming to undertake their study. The progress in the investigation of living beings had to wait for the arrival on the scene of an English physician, William Harvey.

Educated at Padua and Cambridge

William Harvey was born on April 1, 1578 in Folkotone, England, the eldest son of Thomas Harvey, a yeoman farmer and landowner. After attending King's School, Canterbury, Harvey entered Calus College, Cambridge University. He studied arts and medicine during 1593-1599.

When Harvey was 20, he proceeded to the University of Padua, the then centre for medical instruction in Europe, where he studied under the celebrated anatomist Fabria and the Aristotelian philosopher Ceremonini.

In 1602, Harvey was awarded a doctorate in medicine at Padua and at Cambridge. Harvey returned to England the same year and set up his practice in London. He was appointed in 1609 physician to St. Bartholomew's Hospital.

Turbulent career

In 1618, Harvey became physician to King James I, a position which he retained under King Charles I. Over the years, he came to be on increasingly close terms with the King. The latter took interest in Harvey's scientific work and provided him with deer from the royal parks for his investigations.

Harvey accompanied the King on his visit to Scotland on his Scottish campaigns of 1639, 1640 and 1641. Following the outbreak of the Civil War in 1642, Harvey was in constant attendance on the King at Oxford from the Winter of 1642 until he surrendered himself to the Scots in May 1646.

Harvey fled with the court from Oxford back to London, and lived at the residence of his brothers. He resumed his medical practice on a limited scale.

Harvey was elected in 1607, a Fellow of the Royal College of Physicians (FRCP), in whose professional and political affairs he took an active interest for the rest of his life. In 1627, he became one of the seven elect of the college; but after 1630, his duties as royal physician increasingly curtailed his participation in the college activities.

Throughout his life, whether at London, at Oxford, or on his various travels, ***Harvey was a painstaking observer of animal life in all its forms.***

Research programme

The earliest display of his scientific work comes from his anatomical lecture notes written in 1616; with these notes, he conducted the anatomical demonstrations for the College of Physicians. These demonstrations were annually repeated at the *Lumelian lectures* for nine successive years. ***It was only after this long probation that he ventured to give his discoveries to the world.***

From these notes, it appears that he contemplated a vast research programme on the motions of the heart, respiration, animal locomotion and generation, the functions of the brain and various other topics; but only his work on the heart and on generation of animals actually came out in print in Latin in 1628.

It was a poorly printed 72-page book *De motu cordis* done by an obscure printer in Frankfurt. ***Harvey arranged printing in Europe, in order to avoid***

trouble in England; for he realised that his ideas flauted the conventional teaching about the heart. Many of his notes were lost when his rooms at White Hall were ransacked in 1642; the rest were destroyed together with his new library by the Great Fire.

His major discovery

By his discovery of the circulation of the blood, ***Harvey revolutionised physiology,*** which had been based on the assumptions about how materials flow through the blood vessels. More importantly, ***his work was one of the major triumphs for modern science,*** as it helped to generate the enthusiasm for science that came to dominate Europe during the second half of the seventeenth century.

Harvey proved his theory on the pathway of blood flow and that ***blood is impelled mechanically by a pumplike action of the heart. He measured the amount of blood in any given unit of time,*** which was one of the first applications of quantitative methods in biology.

Harvey's doctrine of the primacy of the blood, which he had already begun to develop when he wrote his lecture notes in 1616, can be summarised as follows:

All other parts of the body originally receive life from the blood, and thereafter the blood continues to be the main repository of heart and vitality; its constant circulation through the body serves not only to nourish all other parts but also to sustain their heat, spirits, conceptive powers, sensibility and contractility. Thus the ***blood alone is truly alive per se, and the rest of the body is an appendage that serves the blood and lives by virtue of it*** (*The Dictionary of Scientific Biography,* Vol. IV).

Endowments

In 1651, Harvey donated buildings and a library to the Royal College of Physicians. However, he refused to become its president. He further gave in 1656 an endowment to pay a librarian and for an annual oration, which is held till today in his honour.

Harvey, besides life sciences, had interest in literature and philosophy: among his friends were Francis Bacon, Charles Scarburgh and John Aubrey.

Declining years

In his later years, he suffered from gout and kidney stones; he is reported to have survived an attempt in 1652 to end his life by an overdose of laudanum. He died of a stroke on June 3, 1657. His body was 'lapt in lead' and was buried in Hempstead Church.

The style of scientific monograph

The first English edition of Harvey's 72-page book *On the Motions of the Heart and Blood* appeared in 1653. ***It was a landmark in the field of science,*** for it constitutes *the earliest record we possess of a really scientific investigation in biology based on* systematic observation and experiment.

"Although written 350 years ago, the work is essentially modern in tone and method. ***It is, in fact, the precursor and prototype of the scientific 'monograph' of our own day,*** and stands favourable comparison with the best masterpieces of recent times.

In this treatise, Harvey established absolutely the fact of the circulation of the blood, and the fact that the heart was the propulsive agent in the movement. But he was unable, from the want of a microscope, to indicate the precise path along which the blood travelled from the terminal arteries to the commencing veins.

He erroneously conjectured that the blood percolated the organs and tissues as water percolates the earth and produces springs and rivulets." (*Nature,* March 26, 1998.)

ANTONY VON LEEUWENHOEK (1632-1723)

HUNTER FOR MICROBES

Microscopy was his spare time interest. The only knowledge he had was of the Bible in Dutch language. Contrary to education today, education in his days was to learn what was in the books and what the great men of that time had said. Anyone who defied it was seriously punished. ***For the reason that Servitus dissected a human body, he was shot dead.*** When Galelio (1564-1642) tried to explain that earth revolves round the sun, he was imprisoned till his death.

Early years

Antony von Leeuwenhoek was born in Delft, Holland, on October 24, 1932. He left school at the age of 16 due to straitened family conditions and went to Amsterdam where ***he worked as a salesman in a general store for about five years.*** Then he returned to his native place, got married and opened his own general store. Later, for about 20 years he worked for the local municipality in a number of responsible positions. He also worked as a gateman in the same municipality.

Part-time reasearch

His researches, without any formal training in science, were made with the aid of microscopes developed by himself. His hobby was to examine small things he could lay his hands on, employing using a magnifying glass used in cloth trade for examining the fibers.

During his 20 years of part-time research, he learnt the art of grinding glasses, extraction of metal from ores and blacksmith works. Having learnt all these, he built his microscope by arranging these magnifying glasses in a hollow metal tube. He created the best microscope in the country and still he used to feel dissatisfied, always striving to make a better one.

The 'single pebble-lens', which he made, is still with the university museum at Urecht. It has an astonishing magnifying power of 270 and with it, detail can be got down to 1.5 microns. He seems to have made more powerful microscopes than this. All this he could accomplish with no formal education in science, no guide, no laboratory or equipment.

Leeuwenhoek used to observe all the tiny things such as the hair on the skin, muscle fibre of shark, lens of a bull's eye, elk hair, brain of a housefly, the pattern of ovules and seeds of plants, stings of the flea and legs of insects. These were his preliminary observations with his microscope. He was so cynical that he used to observe his specimens for months before noting them on paper. He strongly believed that a scientist, before coming out with something, should be confident of his finding.

In the seventeenth century, ***a group of scientists established the 'Invisible College',*** which strongly came out with the decision that they believe in what they see and not in what great men like Aristotle and Pope have said. Robert Boyle and Issac Newton were among the members of this school.

This school then became the Royal Society of England, when Charles II came to power. It was this society which recognised Leeuwenhoek. Reginer de Graff, a member of this society who investigated into the anatomy of the human uterus, was attracted towards the work of Leeuwenhoek who was asked to send his findings. The society was stunned, when Leeuwenhoek wrote to them his findings and this correspondence went on for about 50 years.

Study of rain drops

One of the most exciting works of Leeuwenhoek was his observation of rain drops. He became more curious when he found microorganisms in these samples collected from the garden and from the roof of his house. He felt they were the smallest organisms present on earth. ***He wondered how such small living creatures were present in rainwater*** and for sometime believed that they were created by God and sent along with rain.

At the same time he came to know that 'cells arise from preexisting cells', as declared by Virchow (1821-1902). To get a clear idea, ***he observed raindrops coming directly from the sky and those from the garden. He found microorganisms only in the latter sample.***

Microbial culture

Thus he concluded that these organisms originate on earth and are not created by a vital force as was believed until then. He constantly observed the water collected directly from the sky for a number of days. In about four days, he found small creatures in this water too! This was a great debacle for the earlier existing theories. He continued to observe the water in wells, lakes and canals and found these organisms in all these samples.

Besides this, he also found how to grow microorganisms by adding organic matter to rain water collected directly from the sky. Thus ***he laid the foundation for 'microbial culture'*** methods, which are still in practice today. He estimated that at least 27 lakhs of microorganisms were present in a single drop of water. These observations made the Royal Society of England dumbfounded.

A proud day

The childish nature of Leeuwenhoek came out when the Royal Society requested him to part with his microscope or the art to build such a microscope for which he was reluctant. He bluntly replied to the society that he would never part with his microscope or reveal the way to construct them, but would allow a team of Royal Society to look at the microorganisms under his microscopes. He never allowed anybody to touch his microscopes.

Owing to the blunt reply of Leeuwenhoek, the Royal Society appointed Robert Hooke and Nehemiah Grew to build microscopes. On November 15, 1677, Hooke presented the microscope to the society and ***the members were amazed when they were able to see tiny creatures in water samples as predicted by Leeuwenhoek. This was a proud day for Leeuwenhoek***. He was elected FRS. As a token of this he replied, "I will serve the society faithfully until my death". He bequeathed to the Royal Society a cabinet containing 26 instruments and extra lenses.

Other discoveries

i. They include the examination of tartar on his teeth. This he did casually and was surprised to see a good number of microbes even in this sample. These microbes were in the shape of 'rods' and described them as ***'Flora of the mouth'. He then examined the tartar of all whom he encountered*** and was able to see the microbes in all of them.

ii. He was the ***first person to observe the sperm in human seminal plasma*** and also differentiated the spermatozoa of man and dog. He observed the capillaries in the tail and fins of fishes and revealed that blood flows in these capillaries as predicted by William Harvey (1578-1657). He discovered the presence of cells in the blood and found that red blood corpuscles were round in mammals and man, and oval in fish and birds. He also detected a small nucleus in the oval cell. He even found the microbes in his stool-samples, both when he was healthy and when he had diarrhoea. Though he observed a number of microbes in both healthy and diseased states of humans, he was not able to conclude that these microbes were responsible for the disease.

iii. Leeuwenhoek also studied yeast, protozoa, hydra giving buds, life cycle of flea, alphids and the way they reproduced, found that 'Cochineal', used as a dye, was composed of the bodies of insects and discovered the sheath surrounding nerve fibers.

iv. He observed how nature regulates the birth of new individuals. In the uterus of the unio he found thousands of embryos and queried, if each unio is able to produce such a large number of babies, the rivers and lakes should be full of these animals. But it is not so. Later he found that the microbes in the water attack the just-born unios and eat them. He reasoned that this was the cause for the controlled population of small unios in nature. He felt that one species depends on the other for its food and it is nature's secret of maintaining an optimal number of a species.

Effect of hot tea

Leeuwenhoek had the habit of drinking hot tea. Once he happened to observe his tartar as soon as he had the tea. He was at dismay when he was not able to find any microbes in this sample. Then he agan observed his tartar prior to taking tea and found these microbes. He was very happy to conclude that microbes died in the earlier sample due to the hot tea and found a way to kill or control the growth of microorganisms by the application of heat. This was the initial step for the 'sterilisation' process which is still in use today.

He propounded a theory to overcome fever. According to his theory, when anyone suffering from fever drinks hot tea, the body of the patient sweats and this helps in cooling the body during its course of evaporation, thereby reducing the fever. This was his personal experience. He followed this principle whenever he was suffering from fever.

Royal visits

Among the dignitaries who personally came and congratulated Leeuwenhoek were the then King of Russia, and the then Queen of England. They also observed the specimens kept under his microscope.

At the age of 85 when his friends requested him to take rest, he commented "a fruit at the ripe stage serves better than the unripe", as he considered this age was his right time to do something more.

He died on August 26, 1723. Even in his deathbed, he called his friend Hugh Vliat and asked him to post his letter about his findings to the Royal Society.

Albrecht von Haller (1708-1777)

Anatomist turned science writer

Rejecting the earlier notions of the self-regulation of the heart, Haller instead explained the regular activity of the heart by the pronounced irritability of its muscles, stimulated by the filling of each section with blood. His thesis on the mechanical automatic working of the heart has been hailed as a landmark.

Early education

Haller was born on October 16, 1708 at Bern, Switzerland. His mother died when he was a child. After receiving early education at home, he attended the local school till 1722, when his father died. During 1722-23, he lived in Biel where his step-uncle took interest in his studies. He was precocious and showed a flair for languages.

In 1724, Haller proceeded to Tubingen where he learned anatomy and botany from John Duvernoy. From there, he went to Leiden, Holland, being drawn by the fame of Herman Boerhaave (1668-1738). He studied anatomy and surgery and graduated at the age of 18 with an M.D.

During the spring and summer of 1728, he studied higher mathematics with John Bernoulli (1667-1748) at Basel. He then started on a botanical collection, which was to form the basis of his massive work on the Alpine flora.

His career

In 1729, ***Haller returned to Bern to practice medicine and conduct***

independent anatomical investigations. But he could not get paid employment. He worked as a private tutor. Finally in 1736, he was selected professor of medicine at the University of Gottingen. He served in this post till 1753, when he returned to Bern in the hope of building a political career. He was director of the Bern salt works from 1758 to 1764, He got interested in public service and was active both in political and administrative affairs. He worked for the school system and the welfare of orphans.

Haller did not have a good health record; he suffered constantly from headache and was plagued by gout, stomach and kidney ailments. All these drove him to despondency, which he tried to overcome by taking opium. He later published a study of his illness. His contemporaries found him amicable and pleasant but at times suddenly irritable and capricious. Some of his scientific work raised up controversies and in turn brought him much trouble and distress, for he was unable to bear another's error in silence. He died at Bern at the age of 69 (December 12, 1777).

Work in anatomy and physiology

As an anatomist, Haller drew on Boerhaave's theory of fibres as the basic structural element of the body. He developed a general theory of cell tissue, which he named tela cellulose: this is also the basis for the tendons, ligaments, cartilage and bones.

He was the first to correlate defined functions of muscle and nerve fibers with specific structures. Haller's investigations of monsters and the study of deformities such as a pair of premature twins, joined at the chest and upper abdomen gave evidence that the anima did not reside in the blood, which was the view held previously.

Haller considered anatomy and physiology as one unit. So his anatomical observations were with studies of physiological function, particularly the heart.

Haller's studies of the structure of the blood vessels, though incomplete, ***are similarly pioneering.*** Haller turned to mechanics from his earlier preparation in mathematics, to define the role of the motion of the blood in the production of heat, which he attributed to the friction produced by the blood corpuscles rubbing against each other and against the walls of the blood vessels.

Anatomical discoveries

Since 1729, he carried on independent research in anatomy: the structure of the human diaphragm, the role of the intercoastal muscles in respiration, the function of cerebellum and the cause of cardiac activity. He asserted, with experimental evidence, that the pleural cavity contains no air, which is ***a significant discovery in understanding the biology of breathing.***

Haller set an important example for his followers in drawing conclusions from experiments rather than analogies, though some of his interpretations were erroneous. Nevertheless, ***his ideas concerning irritability and sensibility became the basis of a medical system.***

He applied his experimental methods to studies on embryological development, to which he came, through his investigations of the human gonads. He devised a numerical method to demonstrate the rate of growth of the fetal body parts. He showed that fetal growth is relatively rapid in its earlier stages but the tempo gradually decreases. These observations were entirely new and remain his original anatomical discovery.

As a science writer

Haller's preoccupation with anatomy and physiology did not turn his attention away from the science of botany. He set himself to create a complete encyclopedia of Swiss flora (1768). This major treatise, brought him universal recognition: several plants were named after him. He did not neglect his herbarium: his plant collections have been maintained in the National museums at Gottingen and Paris.

Along with his original scientific work, ***he took zealously to his literary career as a science writer.*** His numerous book reviews published in the *Gottingen Times,* of which he was the editor (1747-1753) accounted for his immense knowledge of scientific literature. He published extensively: a two-volume book in botany, transcripts of Boerhaave's lectures with his annotations which enjoyed popularity as a textbook of physiology, an eight-volume handbook of physiology, a comprehensive Bibliotheca medica, listing more than 50,000 titles, which cited historically interesting relationships between the authors and their works and three philosophical romances.

(*Dictionary of Scientific Biography,* Ch. Scribners Sons, New York.)

FRIEDRICH WOEHLER (1800-1882)

PIONEER OF BIOCHEMISTRY

Woehler's work was more important in opening up whole new dimensions of biochemistry, stimulating work on the chemistry of digestion, respiration, growth and reproduction.

Passion for practical chemistry

Friedrich Woehler was born on July 31,1800 in Frankfrut-am-Main, the son of a veterinary surgeon. After attending the local gymnasium from 1814 to 1820, he studied medicine at the University of Marburg and Heidelberg. Woehler qualified for the M.D. degree specialising in gynecology, but he never practiced.

From childhood Woehler had a passion for practical chemistry. The year (1823-24) he spent with Berzelius in the analysis of minerals indeed shaped his career. The friendship with Jacob Berzelius (1779-1848) was strengthened through voluminous correspondence.

Professor of chemistry at Gottingen

His teaching career began at an industrial school in Berlin (1825-31) and in Kassel (1831-36). He finally settled at Gottingen as professor of chemistry, which chair he held for 46 years.

Both as a teacher and experimenter, Woehler was outstanding. It has been recorded that ***about 8000 students passed through his hands.***

Preparation of artificial urea

Woehler's most famous discovery occurred in 1828, when he synthesised crystals of urea while evaporating a solution of ammonium cyanate. He wrote excitedly to Berzelius, "I must tell you I can prepare urea without requiring a kidney of an animal, either man or dog". The significance of his achievement was that urea is an organic substance; which, it was hitherto thought, only a living organism could synthesise. If the constituents of a living body can be put together in a laboratory like common salt or sulphuric acid, then there is apparently nothing left to distinguish the living from the non-living. For this reason, Woehler's work is frequently cited as marking the death of vitalism, although at the time, Woehler was probably more concerned with the chemical reactions involved. However, Wohler's contemporaries did not see it as having that significance. Just because one substance had been synthesised, no one was prepared to claim that all organic substances could be so created.

Justus von Liebig (1803-1873), who knew Wohler's work well and collaborated with him over a long period of time, was a vitalist and vitalism was too complex and deep-seated an idea to disappear as a result of one experiment. Woehler's work was more important in opening up ***whole new dimensions of biochemistry, stimulating work on the chemistry of digestion, respiration, growth and reproduction.***

Extraction of aluminum and beryllium

Woehler made other contributions to organic chemistry. In 1832, in collaboration with Liebig, ***he showed that the benzoyl radical (C_6H_5CO) could enter unchanged in a series of compounds:*** the hydride, chloride, cyanide and oxide. Thus organic chemistry became, for a time, the chemistry of compound radicals. With the theories of Berzelius, this approach led to a great increase in the knowledge of organic compounds without a corresponding understanding of their chemistry.

In fact, ***organic chemistry became so confusing that Wohler returned to inorganic chemistry.*** In later years, he tended to concentrate on the chemistry of metals, in particular the production of pure samples of some of the less common metals. He succeeded, at great expense, in obtaining pure aluminum (1827) and berylium (1828).

Collaboration with Liebig

With Liebig, he was partly responsible for the discovery of isomerism. In 1823, while working in the laboratory of Berzelius, he prepared silver cynate; at the same time, Liebig produced silver fulminate, a compound with very different properties. To their surprise, they found both compounds had identical formulae. ***Berzelius named the phenomenon 'isomerism'.***

One of the final achievements of Woehler was the creation of one of the first greatest teaching laboratories of Europe at Gottingen. From the 1830s nearly all creative chemists of the 19th century spent some time at Gottingen; students came not just from the continent but also from America and Britain. The tradition continued until the rise of Hitler.

Woehler's most distinguished student was Hermann Kolbe; and the latter and Liebig remained his life-long friends. He continued to be a teacher, taking interest in what went on in his laboratory even in old age, till the end (September 23, 1882).

Charles Darwin 1809-1882

Who revolutionised biology

Darwin himself did not become deeply involved in the defence of his theory leaving that to others, notably Thomas Huxley. One of the book's greatest effects was the spur it gave to biological research.

Flair for geology and natural history

Charles Darwin was born on February 12, 1809 at Shrewsbury, England. His mother died when he was eight. He was reared by his elder sisters. He entered the local school where he was taught nothing but classics, ancient history and geography. But ***the grounding in classics helped him afterwards to think so devastatingly straight and to use English language with precision***. His headmaster rebuked him publicly as a dull student. So, he was made to quit Shrewsbury school in 1825 and was sent to Edinburgh University to study medicine, which he did until 1827.

But Darwin found the lectures on materia medica, anatomy and pharmacy unbearably dull. He turned away from a career in medicine, after witnessing several operations performed without anesthesia. At the same time, he began to show interest in geology and natural history.

Faced with the problems of starting him on a new line, his father sent him to Cambridge University (1828) to study theology to enter the Church of England as a clergyman. He had no inclination for the ministry; instead he made the acquaintance of scientists Adam Sedgwick and John Henslow, who turned his mind in the direction of natural history. Darwin took a poor degree at Cambridge in 1831, when he was back home.

Turning point

A turning point in his life occurred through the friendship of Henslow. Darwin was offered on the recommendation of Henslow an unpaid position as naturalist in the HMS Beagle, to study the Pacific coast of South America and some Pacific islands. The Beagle left on December 27, 1831 and returned on October 2, 1836. Darwin spent 535 days at sea and 1200 days on land. Darwin sailed with no formal scientific training.

Darwin kept his field observation in notebooks (600 quartos) with the specimens listed serially and their place and time of collection documented. ***He returned a true man of science*** knowing the importance of evidence, to propose that species had not always been as they were since the creation but had undergone change.

He also had doubts of the value of scriptures as a trustworthy guide to the history of the earth and the man, he gradually became an agnostic. How he dealt with these experiences and what they led to, built up into a process of epoch-making importance in the history of thought.

The Zoology of the Voyage of the Beagle was published in five volumes, from 1839 to 1843. Specialist systematists wrote on fossil and living mammals, birds and fish. Darwin edited the work and contributed habits and ranges of the animals and geological notes on the fossils. Darwin's contribution falls into three main groups: evolution and natural selection, botany and geology, and the origin of species.

Meticulous research

Established biological thinking held that all species were unchanging and specifically created. However, along the Pacific Coast of South America, Darwin observed that closely related species often lived in adjacent geographic areas. When he found at a given location fossils of extinct species that were similar to living species, he began to question why similar species existed in successive geologic time periods.

Darwin collected many geologic and biologic specimens, studied many fossils, and made observations of the form, number, diversity and living habits of different forms of life. From his meticulous research, he arrived at the idea that species descend with modifications, from other species. At a stroke, he established humans just as a species of animal, rather than a lordly creature created in God's own image, which most people at the time believed. It was a shift in self-perception for many that was as humbling as troublesome.

Just as Copernicus cast the earth out from the centre of the universe, Darwin displaced human species as the epicentre of the natural world. It took a century of further research work, with the discovery of DNA, to pinpoint the genetic mechanism of change (*Scientific American*, 2009).

Publication of the Origin of Species

Darwin published nothing on this subject for over 20 years, during which he completed the publication of the journals of his voyage. In 1856, Darwin began to write the history of evolution by natural selection, but, ***when he was halfway through, he was stunned in 1858 to receive from Alfred Wallace (1823-1913) a summary of the work he had been doing for 21 years***. However, friends arranged for the two men to present a joint paper before the Linnean Society, London. On November 24, 1859, Darwin at last published his work. *The Origin of Species*.

The publication of this book secured worldwide attention and aroused impassioned controversy. It upset many established patterns of thought, ***contradicted firmly held religious tenets and brought into focus the concept that humans are one species among many that had evolved from a more primitive one***. Controversies on the theory raged all over England, Europe and the U.S.A.

Darwin himself did not become deeply involved in the defence of his theory, leaving that to others, notably Thomas Huxley (1825-1895). One of the book's greatest effects was the spur it gave to biological research. Darwin expanded his ideas in the books that followed: *'The Descent of Man'* (1871) and *'The Expression of the Emotions in Man and Animals'* (1872).

Biological findings

Darwin's work in biology gave him a lasting place in natural history. On the Galapagos Islands, Darwin found a dazzling array of animal life and found that related but different species lived on different islands, even though all these islands were very similar in geological, climatic and other physical conditions.

For example, he found an array of gold finches with beaks ranging from large and powerful to small or fine. He correlated these differences not with physical conditions on the islands but with the bird's feeding habits.

Those with powerful beaks ate large seeds; those with smaller beaks ate small seeds; those with fine beaks primarily feed on insects. So he postulated that each finch was particularly suited to the food that was available in its environment — this phenomenon is now labelled 'adaptation'. These birds are today named Darwin's finches.

Geological research

Darwin made important discoveries, many of which later provided ***geological support for his theory of evolution.*** He observed that sedimentary rock crystallises when metamorphosed by overlying rock and saw evidence of how volcanoes and earthquakes change the lay of the land, uplifting some areas and sinking others. By these geological observations, Darwin established that things in nature change with time. He published his conclusions in three books: *Coral Reefs (1842), Volcanic Islands (1844),* and *Geological Observations on South America (1846).*

Since the HMS Beagle voyage, Darwin was plagued by chronic ill health caused by frequent insects bites in Argentina. He found himself handicapped by lassitude and nausea. So he settled (1842) in the village of Downe, 15 miles from London.

He restricted his activities to a daily routine of four hours' work, walks in his estate and periods of rest on a sofa while reading a novel, listening to the piano and going to bed early. Four of his sons became prominent scientists.

The other naturalist

The only other figure regarded as important as Darwin for our understanding of the evolution of life on Earth is the ***unknown Russian Scientist*** Vladimir Vernadsky. His seminal text *The Biosphere,* written in 1926, is now available in English. He was posthumously accepted by the Russian scientific establishment: he appeared on stamps and coins, had a mineral, several mountains, a street in Moscow and even a crater on the Moon named after him. (*New Scientist,* March 14, 1998.)

Honours

Darwin was elected FRS at the age of 29. Three foreign universities awarded him honorary doctorates; 57 foreign learned societies elected him to honorary membership, but the French Academy of Sciences did so only in 1878.

He never received any recognition from the British Government, except a burial in Westminster Abbey on April 19, 1882.

(*Collins Biographical Dictionary of Scientists,* 1993 and *Grolier Academic Encyclopedia,* 1994)

ALFRED WALLACE (1823-1913)

A DEDICATED NATURALIST

Most people know Alfred Wallace as the other scientist who conceived the theory of natural selection's contribution to evolution theory, but was forever eclipsed by Charles Darwin. From recent studies he emerges as a foremost scientist in his own right, a pioneer in zoogeography.

Early Life

Alfred Wallace was born on January 8, 1832 at Usk (England) and had a rudimentary education. He earned his livelihood doing a variety of jobs, which included teaching in a school at Leicester (1844).

He was deeply influenced by the books written by Alexander von Humboldt (1769-1859) and Charles Darwin (1809-1882). He came to the conclusion that in tropical America, he would find the materials for working on the origin of species.

With H.W. Bates, an entomologist, he went in 1848 on an expedition to the Amazon. Unfortunately, most of the specimens collected were lost during the return voyage, as the ship was destroyed by fire.

Restarting the quest

Wallace restarted his quest as a naturalist at the Malay Archipelago where he spent over eight years (1854-1862). His observations came out as an essay (written in Sarawak 1855). 'On the law, which has regulated the introduction of new species' was a significant pre-Darwinian contribution to the theory of evolution. It attracted little attention, but Darwin recognised its similarity.

After studying the *Essay on the principle of population* by Malthus and considering the arguments postulated, Wallace elaborated the above-cited essay. It also included his views on natural selection, in an essay *On the tendency of varieties to depart indefinitely from original type.* This he sent to Darwin in 1858 for his comments, ***"I never saw a striking coincidence: if Wallace had my MS sketch written out in 1842, he could not have made a better short extract!".***

Both papers were presented at the at the same meeting of the Linnean Society, London on July 1, 1858. Wallace's diffidence led him to a subordinate role in the history of science. It was Wallace who first synthesised all available evidence in his *magnum opus, Geographical Distribution of Animals* (1876).

Wallace divided the Malay archipelago into two distinct zoological regions by the strait between Bali and Lombok, with oriental fauna on the western side and Australian fauna on the eastern side.

He wrote several books, which excited the admiration of naturalists: *Malay Archipelago* (1869), *Island Life* (1880), *Darwinism* (1889). ***He defied Darwin's theory by arguing that natural selection could not explain the complexity of the human mind.*** He was an eloquent essayist and published his views on spiritualism which lost him some support, but not the respect.

Darwin's tribute

Honors came to Wallace only late in his life: FRS (1893) OM (1910). Darwin recognised the striking coincidence of Wallace's contributing to his theory, which he expressed in his tribute. "You are the only man I ever heard who personally does himself an injustice, and never demands justice". (*Biographical Dictionary of Scientists,* Harper Collins Publishers, 1994)

Ivan Pavlov (1849-1936)

Pioneer in Physiology

The basic pattern, evolved from the unconditioned and conditioned response, had led to a remarkable proliferation of concepts, which have pervaded many parts of physiology other than that of the nervous and digestive systems.

Study of diverse subjects

Ivan Pavlov was born on September 26 1849, at Ryazan, Russia, the son of a poor parish priest. ***From his father, he imbibed a love for labour and learning.*** Around the age of 9, he met with a fall, which affected his general health and delayed his going to school.

Pavlov was sent at the age of 11 to the church school at Ryazan. In 1864 he entered the local seminary, studying religion, classical languages and philosophy. He also developed an interest in science. He did not complete his studies; but in 1870, he gained admission to the University of St. Petersburg, choosing animal physiology as his major and chemistry as his minor. He studied organic chemistry under Mendeleev (1834-1907), but the lectures of Ilya Tslon made a lasting impression on him. It was in Tslon's laboratory that Pavlov got the introduction to scientific investigation. His work on the nerves that govern the pancreas (1875) won him a gold medal.

Career in teaching and research

On graduation from St. Petersburg in 1875, he entered the Military Medical

Academy to study theoretical medicine. In 1877, he published his first paper 'Experimental data concerning the accommodating mechanism of the blood vessels'. From 1878 to 1890, he headed the physiology laboratory of Sergel Botkins and conducted investigations on the physiology of blood circulation and of digestion. He received the M.D. degree in 1879. Pavlov completed his thesis, 'The centrifugal nerves of the heart', for which he was also awarded the doctorate in 1883.

The following year, he was appointed lecturer in physiology at the academy. In 1890, he was made professor in the department of pharmacology. Five years later, he accepted the chair of physiology, which he held till 1925.

In his last article, 'The Conditioned Reflex' (1934), he postulated his theory of the two signalling system which differentiated the animal nervous system from that of man.

Research in physiology

Pavlov's research covers three phases: the physiology of the circulation of the blood (1874-1888), the physiology of digestion (1879-1897), and the physiology of the brain (1902-1936).

In the first phase, Pavlov developed operative surgical techniques that enabled him to perform experiments on unanesthesised animals without inflicting much pain. He showed that cardiac function is governed by four nerves which respectively inhibit, accelerate, weaken and intensify it.

In the second phase, Pavlov concentrated on the nerves directing the digestive glands. He discovered the secretion nerves of the pancreas (1888) and the nerves concerning the secretion of the gastric glands (1889). Pavlov and his students produced a considerable amount of accurate data on the workings of the gastrointestinal tract, which were published as 'Lectures on the work of the principal digestive glands' in 1897.

During the final phase of his scientific career, Pavlov turned to the physiology of behaviour. By the beginning of the 20th century, many physiologists, zoologists and psychologists had already undertaken experiments to study the function of the brain. Pavlov drew upon Darwin's theory of evolution and the theory of brain reflexes of his contemporary Sechenov (1829-1905).

Fundamental facts and concepts

Pavlov investigated the activity of the cortex and the cerebral hemispheres, basing his work on fundamental facts and concepts of the physiology of the nervous system. He chose to approach these areas through studying salivary glands, because their activity could be subjected to quantitative measurement.

Food placed in a dog's mouth induces gastric juice flow, and 'unconditioned' response, according to Pavlov. If, however, a bell is rung regularly before the food is presented, the response, which may eventually take place in the absence of food, is 'conditioned'. ***This basic pattern had led to a remarkable proliferation of concepts, which have pervaded many parts of physiology other than that of the nervous and digestive systems***. Pavlov presented his theory on the conditioned reflex in a paper presented to the Fourteenth International Medical Congress held in Madrid (1903).

Impact of experimental method

Pavlov introduced into the laboratory the method of long-term experiment, not only as a technique but also as a way of thinking. This gave birth to a new era in physiology of digestion and led to new work and concepts, especially in experimental surgery and in the physiology of the brain. Pavlov said ***"Science moves in spurts, depending on progress made in its methods. With each step forward, we rise, so to speak, to a higher step from which a wider horizon opens to us, with subjects previously unseen".***

Honours

Pavlov's scientific contributions received world-wide recognition. For his work on digestion he received the ***Nobel Prize in Physiology or Medicine*** (1904). He was elected in 1907 an Academician of the Russian Academy of Sciences (1907).

Pavlov organised a number of major research centres — the Institute of Physiology (that now bears his name), the biological station at Koltushy (now Pavlov). The Russian Physiological Society has been named after him.

Pavlov did not welcome the February Revolution of 1917; he publicly stated "If that which the Bolsheviks are doing with Russia is an experiment, for it I should regret even giving a frog". ***Despite his hostility, the Soviet authorities assured Pavlov of continuing support for his work and special privileges.*** By 1935, Pavlov became reconciled to the Communist system, remarking that the government, too, is an experimenter but in an immeasurably higher category. Pavlov became seriously ill in 1935. He died on February 27, 1936 in Moscow.

(*Dictionary of Scientific Biography,* Ch. Scribner's Sons, New York.)

Paul Ehrlich (1854-1915)

The founder of chemotherapy

Among medical scientists of his generation, Ehrlich was probably the most original and successful. The fruitfulness of his concepts initiated advances in all fields of biomedical research. Hematology became a recognised discipline through his pioneering studies of dye reactions on red and white blood cells.

Fascination towards chemistry

Paul Ehrlich was born on March 14, 1854 in a comfortable household in a country town, Strehlen near Breslau, Germany. He was the only son and fourth child of a respected Jewish distiller and inn-keeper.

At the age of 10, he went to the St. Maria Gymnasium in Breslau. ***He disliked*** all examinations. After matriculating in 1872, he studied for his Physikum at Strasbourg, which largely determined the course of his career. ***He became fascinated with the subject of chemistry and began making excellent histological preparations with his own modifications of aniline dyes.*** Ehrlich studied medicine in the Universities of Breslau, Strasbourg and Leipzig, from where he graduated in 1878.

While still a student, he began investigations on the aniline dyes. With their aid, he discovered all the different types of white blood corpuscles.

His career

Ehrlich was appointed on his graduation, chief assistant to the renowned physician Friedrich von Frerichs, at the Charity Hospital in Berlin. He gained lasting insights into diagnostic and therapeutic problems. When Frerichs died suddenly in 1885, he found his research impeded, causing disturbance in his program of work.

In 1888, discovering tubercule bacilli in his sputum, he ended a decade of fruitful association with the clinic and travelled with this young wife to Egypt, where he spent a year. In 1889, he returned to Berlin apparently cured of pulmonary tuberculosis by Koch's newly discovered tuberculin treatment and never had a recurrence.

Ehrlich then set up a small private laboratory and launched a series of fundamental studies in immunity that engaged his attention for many years. He worked for more than three years without salary at the newly founded Institute for Infectious Diseases, of which Robert Koch (1843-1910) had become director. Eventually, this institute was affiliated in 1906 to the Georg Speyer Institute for Chemotheraphy; he was the director of these establishments till his death (9 August 20, 1915). On the founding of the University of Frankfurt, he was appointed a professor (1914).

His ingenuity with dye stuffs

Ehrlich explored the avidity of living tissues for certain dyes. In a remarkable monograph (1885), he reported his investigations into the distribution of oxygen in animal tissues and organs. Using two vital staining dyes, alizarin blue and indophenol blue, he demonstrated that while living protoplasm in general has potent reducing properties, bodily organs can be classified according to their oxygen acidity. ***He enunciated the fundamental concept underlying his work: chemical affinities govern all biological processes.***

Koch's dedication to problems of infection, his own experiences with tuberculin treatment and Emil von Behruig's discoveries of diphtheria and tetanus antitoxins led Ehrlich to investigate bacterial toxins and antitoxins by methods comparable with those employed in his plant protein studies.

Development of 'Salvarsan'

At the ceremonial opening of the Georg Speyer Institute (September 1906), ***he remarked prophetically about the creation of substances 'in the chemist's retort'*** that would be able to exert their full action exclusively on the parasite harboured within the organism and would represent 'magic bullets', which seek their target of their own accord.

This culminated in his announcement before the Congress for Internal Medicine at Wiesbaden (April 1910) that a synthetic arsenical compound, which he named 'salvarsan' had shown curative properties in rabbit syphilis and also in clinical trials. After successive modifications of this compound, it was first used in the treatment of human syphilis in 1911. ***The science of chemotherapy was born.***

The invention of salvarsan brought Ehrlich four years of both tragedy and triumph. He battled with problems that stemmed from the drug's imperfections, from the complex pathology of syphilis, from human cupidity and malice. His institute overflowed with visiting physicians and would-be patients.

Secret of his success

Ehrlich harnessed brilliant talents. In addition, he possessed the faculty of stereognostically visualising benzene rings and structural chemical formulae; unique virtuosity in 'test-tube chemistry'; ***the capacity to direct several investigators simultaneously, through a system of daily written instructions;*** and the foresight to abandon paths that were unpromising.

Honours

Many honours came his way: ***Nobel Prize for Medicine (1908),*** the Prussian Great Gold Medal for Science (1903), the Liebig Medal (1911), the Freedom of the city Frankfurt (1912), the title Excellenz (1911); the Paul Ehrlich prize was awarded by the Paul Ehrlich Institute as a living memorial to him.

Sigmund Freud (1856-1939)

Founder of psychoanalysis

Pychoanalysis provides a major tool of research. The application of this new science to the studies of art, anthropology, history and education has greatly changed these fields.

Influence of his mother

Sigmund Freud was born on May 6, 1856, in Freiberg, Moravia Canton, in the erstwhile Czechoslovakia. He was the eldest child in his father's second family. An older stepbrother, with a child about Sigmund's age, was his playmate.

The boy grew up in an ***unusual family structure,*** his mother half way in age between himself and his father. All this caused him confusion; but it sharpened his intellect. He remained his mother's favourite in a family of seven children.

He wrote of himself: "One who has been the indisputable favourite of his mother keeps for life the feeling of a conqueror, the confidence that often induces real success". His unusual degree of ambition lasted well into his middle years.

His father was a wool merchant in Freiberg and when the trade collapsed, the family moved in 1860 to Vienna. The father was often unemployed and so the family was at times on the brink of poverty.

Choice of medicine

Freud attended Sperl Gymnasium in Vienna from the age of 9 to 17, studying Greek and Latin, mathematics, natural sciences and history. He passed his final examination with distinction.

His family recognised him as a superior student from the beginning and made considerable financial sacrifices for his education. Freud met his fiancée Martha in 1882 but a period of five years preceeded their marriage in 1887. Of their six children, Dr. Anna Freud became one of Freud's most famous followers.

Freud first considered studying law but then enrolled in 1875 in the medical school at the University of Vienna, which had become the world's capital of medicine. ***He chose medicine,*** not out of a desire to practice it but ***to study the scientific side of medicine.***

In this ambition, he took Ernst Bruecke, his professor, as a model. The latter was one of the founders of the Helmholtz school of physiologists, the progress of which they accelerated by their own work and teaching. Freud had to spend three more years than was necessary in qualifying for his M.D. degree in 1881. He spent the period (1876-1882) in the Zoological and Anatomical laboratories of Bruecke's Physiological Institute.

Studies in clinical neurology

Freud studied the structure of nerve fibers in living crayfish and the anatomy of the human brain. These studies augured well for a successful research career when he saw that opportunities for advancement in the academic rung of the profession were rare. So he decided to shift from the Institute to the Vienna General hospital and worked from July 1882 to August 1885 as resident staff. He stayed for 14 months in the department of nervous diseases.

Besides, he continued his anatomical research on the human brain, tracing the course of nerve tracts in the medulla oblongata. Freud also began a series of studies in clinical neurology, which was in the tradition of investigation of the patients in the hope that it would lead to a correct diagnosis of behaviour in mental disorders. His later work revolutionised this attitude. He published papers on neurology and brain function including cerebral palsy and aphasia, which were widely acclaimed.

During the last part of his residency, Freud was deputed for four months to study under the neurologist J.M. Charcoat (1825-1893) in Paris. He saw in Charcoat's demonstration the psychological origins of hysteria, which led to his development of psychoanalytic approach to mental disorders.

Development of psychoanalysis

Freud returned to Vienna and established himself as a private practitioner. He devoted his efforts to the treatment of hysterical patients with the help of hypnosis. With Joseph Breuer, his friend and mentor, he published *Studies on Hysteria* (1895), which included several chapters and a series of case studies. He coined the term 'psychoanalysis' and his major life work was well under way. Freud began also a unique undertaking, his own self-analysis: as he proceeded, his personality changed.

The result was the publication of his book, *The Interpretation of Dreams?* (1901) where he demonstrated that the dreams of every man serve as a royal road to the understanding of unconscious mental processes. By the turn of the century, Freud had increased his knowledge of the formation of neurotic symptoms other than hysteria. Freud wrote a series of papers, in which he explored the influence of unconscious mental process on virtually every aspect of human behaviour: *The Psychopathology of Everyday life (1901), Leonardo da Vinci* and *Memory of His childhood* (1910), *Totem* and *Taboo* (1912).

He presented his views on infantile sexuality in *Three Essays on the Theory of Sexuality* (1905), a work which initially met violent protest but was gradually accepted by all schools of psychology. During this period, he also published a series of articles dealing with psychoanalysis as a therapy.

Freud soon became ***the intellectual father figure of the psychoanalytic movement.*** The Wednesday meetings of his followers began in 1902 at Freud's house. The first International Psychoanalytic Congress was held in Salzburg in 1908.

Recognition in the U.S.

In 1909, Freud was invited to give five lectures at Clark University, U.S., which was the first official recognition of his new science. Psychoanalysis became entrenched in the medical school curriculum. ***The general increase***

in the prosperity in the U.S. showed up in the marked increase of psychoanalysts.

Several followers of psychoanalytic movement in Europe left — notably Eugen Bleuler, Alfred Adler and Carl Jung, because of personal conflicts with Freud. Earlier, Freud and Breuer had differed concerning their findings with regard to the role of sexual desires in neurosis. The two men parted company, not without some personal animosity. Ironically, Breuer was remembered by later professionals because of his brief collaboration with Freud.

In 1923 Freud developed cancer of the jaw that led to his death (September 24, 1939) 16 years later. From then on, he had 33 operations, causing him a lot of pain. He preferred to remain mentally alert rather than take pain-killing drugs.

Clinical studies on himself

In spite of this major illness, he wrote three books, *Beyond the pleasure principle (1920), The Ego and the Id* (giving his theory of anxiety, 1923) and *Inhibitions, Symptoms and Anxiety* (1926). Freud's home and extensive practice occupied most of his time. He was throughout his career, on the faculty of medicine, without chaired appointment of the University of Vienna. He was acclaimed as a talented writer and essayist.

With the impending danger from the Nazis, Freud was offered foreign asylum in 1936 and 1937, but partly out of his identity with less fortunate Jews, he remained in Vienna till June 1938. He was allowed to go to England, after the British psychoanalyst Ernest Jones managed some complicated diplomatic maneuvering and the payment of a ransom. He died on September 23, 1939 in London.

Freud's personal life has been a subject of interest to admirers and critics. ***When it seemed necessary to advance his science, he exposed himself mercilessly to clinical studies: his own mental functioning was the major subject matter of psychoanalysis.***

Application of psychotherapy

Under his guidance, psychoanalysis provided a major tool of research. Psychiatric treatment currently commands thousands of practitioners all over the world. The application of this new science to the studies of art, anthropology, history and education has greatly changed these fields. He inspired loyalty to this school of thought in a circle of disciples that continues to this day. (*The Dictionary of Scientific Biography,* Vol. 4.)

WILLEM EINTHOVEN (1860-1927)

INVENTOR OF THE ELECTROCARDIOGRAPH (ECG)

Among the physical methods used in clinical medicine, electrocardiography is second in importance only to the discovery of X-rays by Roentgen (1845-1923).

Early Life

Willem Einthoven was born on May 21, 1860 at Semarang (East Indies) where his father practiced as a physician. The latter died in 1866. The family shifted in 1870 to Utrecht, Holland, where Einthoven graduated from the local high school. He qualified in medicine in 1885. He was appointed professor of physiology at the University of Leiden, at the young age of 26.

While a student, Einthoven was active in sports: when he broke his wrist in a fall, he used this incident to publish a study on the 'pronation and supination' of the forearm (1882). ***The main clinical applications of medicine became his interest hereafter.***

Construction of the recoder

In 1895, A.D. Waller (1856-1922) had published the curve for the action current of the heart as deduced from the body of surface, but he was unable to register its true shape.

Einthoven repeated the experiment, adapting the celebrated mirror galvanometer of de Arsonval, he constructed the string galvanometer and applied it to record the electrical activity of the contracting heart muscle. He recorded the curve's true shape, which he named the electrocardiogram.

Einthoven recorded ***the electrocardiograms not from the chest wall but chose both hands and left foot as contact points.*** This gave three possible combinations for contact, which he labelled: I (both hands); II (right hand–left foot); and III (left hand–left foot). The information received from the contacts thus represents the projection of what takes place in the heart.

The few classical papers that he published between 1903 and 1908 touched on the possibilities of using the technique in the diagnosis of diseases of the heart, but ***were concerned mainly with the physical principles.***

Clinical applications

These followed a few years after Einthoven's discoveries. They were treated in the definitive accounts which came from eminent physicians William Osler (1849-1919) and James Herick (1861-1954).

The correlation of electrocardiograms and of clinical data was achieved between 1910 and 1935 by the schools of Thomas Lewis in London and F.N. Wilson in the U.S.

Einthoven extended his technique to cover many forms of heart disease, by adding a second string galvanometer to register heart sounds and murmurs simultaneous to the electrocardiogram. He continued his work to perfect the sensitivity of his string galvanometer, attempting to record nerve impulses at the time of his death (September 28, 1927). This was achieved only later by other workers who used the cathode ray oscilloscope.

Einthoven received the ***Noble Prize for Medicine or Physiology in 1924.***

(*The Dictionary of Scientific Biography,* Vol. IV, Ch. Scribner's Sons, New York.)

Salim Ali (1896-1974)

Bird lover and ecologist

Salim Ali generated his own income, from the sale of his immensely readable books and many awards. But he kept very little for himself, as his needs were few. He supported needy students pursuing research from his own funds. Besides, he spent large sums on postage in corresponding with his innumerable friends all over the world or in voicing his anguish for protecting the environment. Like his birds, he worked by day and ate very little.

An incident that shaped his career

Salim Ali was born in Bombay in 1896 as the youngest of an orphaned middle-class family. He grew up under the care of his maternal uncle, a 'shikari'. He presented young Salim with a 500 shot 'Daisy' air gun, more a toy than a gun.

Salim Ali loved to show it off and one day he shot a sparrow, which had a yellow patch on the throat. His main concern was whether this sparrow was lawful meat for a god-fearing young Muslim or not. He consulted his uncle, who, finding it was a different bird, asked Salim Ali to take it to the Bombay Natural History Society (BNHS) for an authentic pronouncement (fatwa). Mr. Millard, the then honorary secretary, identified it as the yellow-throated sparrow and also showed Salim many stuffed birds in the museum's collection. Mr. Millard encouraged him to start a collection of birds, as the best way of learning about them and explained to him on how to make notes on birds.

Early education

Salim attended local schools till he was nine. At 14, he suffered from chronic headaches, because of which he missed a year in school; he managed to scrape through the matriculation examination in 1913.

Salim Ali then joined the first year course in St. Xavier's College, Bombay, to study biology. But the struggle with logarithms and higher algebra gave him little hope of crossing the formidable barrier in the university curriculum.

Providentially, a letter from his brother in Burma offering a partnership in his business firm, arrived. Though it meant the end of his dream of taking up zoology, particularly ornithology as a profession, he accepted the offer to escape from college mathematics! He sailed for Rangoon in September 1914.

Benign influence of Fr. Blatter

Salim Ali spent the next decade in the small town of Tavoy in Lower Burma, busy in trade. In Tavoy began his romance with natural history, he spent his weekends exploring the countrywide and *watching birds.* He was fascinated by the abundance of the woodpecker. So he began a small reference collection.

He was back in Bombay from October 1917 to February 1919, to take a formal course in commercial law and accountancy. This perhaps proved to be the most momentous period in his life, under the benign influence of Father Blatter and the stimulating tutelage of Prof. J.P. Mullan; he completed simultaneously B.A. in zoology at St. Xavier's College, Bombay. ***Fr. Blatter soon discovered where Salim Ali's interest lay*** and was instrumental in his appointment later as a guide lecturer in the Natural History section of the Prince of Wales Museum.

Salim's introduction to big game came really, after the diversification of their company's declining mining business into leasing government forests for supplying sleepers to the Indian Railways. The timber business, for him, proved to be rewarding: it gave excellent opportunities for camping out and watching forest birds for several days. After three years of unsuccessful business partnership, he returned to Bombay in 1924.

Salim Ali was meanwhile marking time for some congenial natural history assignment by taking up temporary clerical positions, till he joined the post of the Guide Lecturer in 1927. He went to Germany on study leave to work with Professor Stressmann at the Berlin University Zoological Museum (1929-30).

Training in taxonomy

Under the guidance of Stressmann, whom he considered his 'guru', he learned the methods of taxonomy with a collection of Burmese birds he had brought with him. On his return from Germany in early 1930, he found himself with expertise in ornithology but unemployed; it was while retreating at Kihim (near Bombay), did the idea of offering the regional ornithological surveys through the BNHS occur to him. So he conducted bird studies of Hyderabad, the then Travancore–Cochin, and South Indian hills (the Niligris, Kotagiri, Palani). The findings provided the basis for his books, *The Birds of Kerala* and *The Book of Indian Birds.*

Thus, for four years till 1934, he still remained without a permanent job and home. When his brother Hamid suggested Dehra Dun for settling down, he accepted it with alacrity. He would have now at his doorstep among the foothills of the Himalayas, all the things he had dreamt of, beautiful forests, magnificent scenery, 'good birding', trekking and 'naturalising'.

Here, he spent five of his happiest years (1934-39), which was cut short by a personal tragedy, the sudden death of his wife in July 1939. It was while living in Dehra Dun that Salim Ali submitted a project to the then Imperial Council of Agricultural Research, for research in economic ornithology.

He had the vision to observe that the impact of birds on agriculture was of a dual nature — "while birds could on the one hand be destructive to cereal crops and orchards, they could on the other be beneficial in controlling the ravages of insect, pests and rodents, since these largely constituted the food of many species". (*The Fall of a Sparrow,* Salim Ali, Oxford University Press Delhi, 1985)

Salim Ali returned in 1939 to Bombay, as only the BNHS offered facilities for continuing his work. For the next 40 years, he conducted birding expeditions to almost every area of the subcontinent, from the Kutch to Almora and Kashmir to the Nilgiris. The Bharatpur waterfowl resort and

Karnala sanctuary, and the preservation of the Silent Valley were due to his untiring efforts.

Awards and distinctions

He participated in international meetings: Ornithological Conference, Upasala (1950), and the WHO Geneva (1959). To show that even such a seemingly futile occupation as bird watching is not entirely barren of rewards if pursued with dedication. He closed the last chapter of *The Fall of a Sparrow* with a checklist of his awards and distinctions, Padma Bushan (1958). Honorary D.Sc. (1958, 1973 and 1978, ***J. Paul Getty International Prize*** (of $50,000 which he donated to BNHS) for Wildlife conservation (1976), ***National Professor in Ornithology*** (1982).

The last of his ten-volume magnum opus, co-authored with S. Dhillon Ripely (Secretary of the Smithsonian Institution, Washington), *Handbook of the Birds of India and Pakistan* was released on November 16, 1974 — four days after his 78th birthday.

Salim Ali's last wish was to found an ornithological institute at Bombay, which was partially met recently. In 1989, the Pondicherry University instituted ***the Salim Ali School of Ecology*** to commemorate this great bird lover and ecologist.

Thus Salim Ali pursued with courage, born out of love of science and his profession, ***when the subject of ornithology received little more attention than an Englishman's pastime***. In the process, he had to remain without a permanent job and a fixed abode till the end of his life (June 1987).

G.N. RAMACHANDRAN (1922-2001)

A PIONEER IN MOLECULAR BIOPHYSICS

His discovery of the triple helical structure of collagen was a fundamental advance in the understanding of the peptide structure in 1955. This was criticised by Crick and Watson who unravelled the helical structure of DNA, the double helix.

Educated at Bangalore and Cambridge

Ramachandran was born on October 8, 1922 in Ernakulam, Kerala state. His father, a professor of mathematics in the Maharaja's College, taught him mathematics at home. So he gained a strong foundation in mathematics.

He stood presidency first in the Intermediate Examination in 1940, again the presidency first rank in 1942 in the B.Sc (Honours) course at St. Josephs College in Triuchirapalli. ***He chose to study chemistry in place of mathematics, which choice helped him in his later research career.*** His teacher, Professor P.E. Subramaniam, stimulated his interest in physics. So when he enrolled for further studies in the Department of Electrical Technology, Indian Institute of Science, Bangalore, he soon realised that his interests lay markedly in physics.

When his request was persistently refused, ***C.V. Raman, the Director, had to intervene and write "I am admitting Ramachandran into my department, as he is a bit too bright to be in yours".***

From 1942 to 1947, he enrolled as a research student under Raman, obtaining the D.Sc from Madras University. He was awarded the prestigious 1851 Exhibition Scholarship, which gave him an opportunity to work in the crystallography group. He was awarded the Ph.D degree in 1949 by the Cambridge University.

His career

On returning home in June 1949, Ramachandran was appointed as Assistant Professor in the Department of Physics of I.I.Sc. He served for three years till October 1952, when he joined the Madras University as the first professor and head of the Department of Physics. He held this post with distinction for 18 years, ***bringing international recognition to the department***.

On account of the obstacles created to the progress of his research work, he resigned his post. He came bank to his alma mater as professor in 1970 at the invitation of the Director, Satish Dhawan, to start a new department of molecular biophysics. He was head of the molecular biophysics until 1984, when he became the Albert Einstein Professor at the Mathematical Philosophy Group at the institute.

He was Fellow of the Indian Academy of Sciences, Bangalore, Fellow of the Indian National Science Academy, Delhi and Fellow of the Royal Society, London. He died on April 07, 2001 at Chennai.

Influence of Raman

On his first meeting, Raman gave him Lord Raleigh's paper and a book dealing with the topic. Ramachandran found the corresponding equations and solutions for the problem in the book. He worked out a rigorous proof: Raman, being immensely pleased sent it for publication. This happened within one week of his joining Raman.

Ramachandran worked mostly on optics and diffraction theory under the supervision of Raman. It involved photoelasticity and thermo-optic behaviour of different solids such as diamond and fused quartz. His research paper contained some of the earliest applications of X-ray diffraction to the study of variations in the structure of crystals.

The other scientists who influenced his thinking were Lawrence Bragg (1890-1978) and Linus Pauling (1901-1994) who had discovered the alpha helical structure of polypeptides. Ramachandran's vision of chemistry was shaped by reading Pauling's book (*Nature of Chemical Bond*) and his articles.

Structure of collagen

Ramachandran's major contribution was the formulation of the triple helical structure of collagen. J.D. Bernal had remarked, in a casual conversation at Madras in 1952, that the structure for collagen was unsatisfactory. This implanted in Ramachandran's mind the concept underlying his future work. ***The coiled structure he proposed was a fundamental advance in the understanding of the polypeptide structure.*** The triple helical structure of collagen was published, in collaboration with Gopinath Kartha in 1955.

Ramachandran plot

This structure was criticised by Crick and Watson (*Nature,* November 1955) who unravelled the helical structure of DNA, the double helix. To meet their criticism of unacceptably short interatomic contact in the proposed structure, Ramachandran came up with a general method for describing the stereochemical criterion for polypeptide structure and proteins. This work culminated in the 'Ramachandran plot', a two-dimensional map for describing all stereochemically possible structures of polypeptides.

Awards

Ramachandran received over a dozen national and international awards:

- CSIR Golden Jubilee Award for Life Science (1991),
- Sir C.V. Raman Medal (1982), J.C. Bose Award (1976),
- Ramanujam Medal (1972), M.N. Saha Medal (1970),
- Watmul Memorial Prize (1964), Bhatnagar Prize (1961).

Index

ENGINEERING
Appleton, Edward 60
Arago, Jean 18
Babbage, Charles 52
Baird, John 25
Callendar, Hugh 75
Carnot, Sadi 45
Cotton, Arthur 67
Diesel, Rudolf 90
Esaki, Leo 94
Faraday, Michael 3
Fulton, Robert 48
Giaever, Ivar 94
Haber, Fritz 78
Haggerty, Patrick 63
Heaviside, Oliver 12
Henry, Joseph 21
Hertz, Heinrich 15
Karman, Theodore von 81
Kelvin, Lord 6
Krupp, Alfred 72
Langmuir, Irving 55
Marconi, Guglielmo 57
Ohain, Hans von 84
Shannon, Claude 103
Taylor, G.I. 35
Terzaghi, Karl 87
Tesla, Nicola 32
Steinmetz, Charles 28
Stephenson, George 43
Watt, James 38
Whittle, Frank 84
Zworykin, Vladimir 100

PHYSICS
Becquerel, Henri 155
Bohr, Niels 174
Boltzmann, Ludwig 145
Born, Max 161
Bragg, Lawrence 158
Bragg, William 158
Cavendish, Henry 113
Chadwick, James 200
Coulomb, Charles 116
Crookes, William 148
Dirac, Paul 188
Doppler, Christian 203
Fermi, Enrico 215
Foucault, Jean 130
Fraunhofer, Joseph 122
Fresnel, August 125
Heisenberg, Werner 185
Helmholtz, Hermann 134
Huygens, Christian 109
Kapitza, Peter 195
Mach, Ernst 140
Millikan, Robert 178
Pauli, Wolfgang 182
Peltier, Jean Charles 128
Planck, Max 165
Raman, C.V. 210
Roentgen, William 207
Rutherford, Ernst 192
Stefan, Josef 137
van der Waals, Johannes 153
Wien, Wilhelm 168
Young, Thomas 119

MATHEMATICS
Abel, Henrik 239
Bessel, Friedrich 236
Boole, George 247
Bradley, James 280
Cantor, Georg 295
Cauchy, Augustin 267
Chandrasekar, S. 312
Descartes, Rene 221
Erdos, Paul 316
Euler, Leonhard 254
Fourier, Joseph 264
Galois, Evariste 244
Gauss, Carl 270
Halley, Edmund 276
Hawking, Stephen 320
Hilbert, David 298
Herschel, John 287
Herschel, William 283
Lagrange, Joseph 257
Laplace, Pierre 260
Leibniz, Gottfried 229
Pascal, Blaise 225
Poincare, Henri 291
Ramachandra, Yasudas 250
Ramanujan, Srinivasa 306
Riemann, Bernhard 273
Wiener, Norbert 302

CHEMISTRY
Arrhenius, Svante 381
Avogadro, Amedeo 344
Berthollet, Claude 337
Berzelius, Jacob 352
Black, Joseph 327
Bunsen, Robert 369
Dalton, John 341
Dulong, Pierre 356
Fourcroy, Antoine 331
Gay-Lussac, Joseph 348
Hodgkin, Dorothy 391
Hofmann, August von 363
Joliot-Curie, Irene 388
Kekule, Friederich 366
Lavoisier, Antoine 334
Liebig, Justus von 360
Mendeleev, Dmitri 373
Perkin, William 375
Ramsay, William 378
Travers, Morris 384

LIFE SCIENCE
Ali, Salim 433
Darwin, Charles 412
Freud, Sigmund 426
Ehrlich, Paul 423
Einthoven, Willem 431
Haller, Albrecht von 406
Harvey, William 397
Leeuwenhoek, Antony von 401
Pavlov, Ivan 419
Ramachandran, G.N. 437
Wallace, Alfred 417
Woehler, Friedrich 409